Second Edition

The Physical Basis of Chemistry

Complementary Science Series

2000/2001

The Physical Basis of Chemistry, 2nd Edition
Warren S. Warren

Physics for Biology and Medicine, 2nd Edition
Paul Davidovits

Descriptive Inorganic Chemistry
J.E. House ▶ Kathleen A. House

Electronics and Communications
Martin Plonus

The Human Genome, 2nd Edition
R. Scott Hawley ▶ Julia Richards ▶ Catherine Mori

1999

Chemistry Connections
Kerry Karukstis ▶ Gerald Van Hecke

Mathematics for Physical Chemistry, 2nd Edition
Robert Mortimer

Introduction to Quantum Mechanics
J.E. House

www.harcourt-ap.com

Second Edition

The Physical
Basis of Chemistry

Warren S. Warren

Department of Chemistry
Princeton University
Princeton, New Jersey

WITHDRAWN

A Harcourt Science and Technology Company

San Diego   San Francisco   New York   Boston   London   Sydney   Tokyo

Academic Press
A Harcourt Science and Technology Company
525 B Street, Suite 1900, San Diego, California 92101-4495, USA
http://www.apnet.com

Academic Press
24-28 Oval Road, London NW1 7DX, UK
http://www.hbuk.co.uk/ap/

Harcourt/Academic Press
A Harcourt Science and Technology Company
200 Wheeler Road, Burlington, MA 01803
http://www.harcourt-ap.com

Library of Congress Catalog Card Number: 99-68993

International Standard Book Number: 0-12-735855-2

PRINTED IN THE UNITED STATES OF AMERICA
00   01   02   03   04   05   EB   9   8   7   6   5   4   3   2   1

Contents

Prefaces

0.1 Preface to the Second Edition

Everything is made up of atoms. That is the key hypothesis. The most important hypothesis in all of biology, for example, is that everything that animals do, atoms do. In other words, there is nothing that living things do that cannot be understood from the point of view that they are made of atoms acting according to the laws of physics. This was not known from the beginning; it took some experimenting and theorizing to accept this hypothesis, but now it is accepted, and it is the most useful theory for producing new ideas in the field of biology.

Richard Feynman (1918–1988)
Nobel Laureate in Physics,1965

It would be a poor thing to be an atom in a universe without physicists, and physicists are made up of atoms. A physicist is an atom's way of knowing about atoms.

George Wald (1906–1997)
Nobel Laureate in Physiology/Medicine, 1967

With all due respect to Feynman (one of the towering figures of twentieth century physics), his quote slightly misses the point: it would be better to say that "everything animals do, *molecules* do". And as Wald certainly knew, physicists are better understood as collections of molecules than as collections of atoms.

Of all of the remarkable scientific achievements of the late twentieth century, none is more spectacular than the transformation of biology into molecular biology and its associated subdisciplines. This transformation occurred only because, time and time again, fundamental advances in theoretical physics drove the development of useful new tools for chemistry. The chemists, in turn, learned how to synthesize and characterize ever more complex molecules, and eventually created a quantitative framework for understanding biology and medicine. We chemists like to describe our field as *the central science*, and indeed it is. Our job as educators is to help students understand the interconnections.

This small book grew from my supplementary lecture notes during the ten years I have taught advanced general chemistry or honors general chemistry at Princeton University. It is mainly intended as a supplement for the more mathematically sophisticated topics in such courses. I have also used parts of it as background for the introductory portions of a junior-level course, and it has been used elsewhere as an introduction to physical chemistry. For example, an introduction to biophysical chemistry or materials science should build on a foundation which is essentially at this level. Most of the students become science or engineering majors, and they have a broad range of interests, but the strongest common denominator is interest in and aptitude for mathematics. My intent is not to force-feed math and physics into the chemistry curriculum. Rather it is to reintroduce just enough to make important results *understandable* (or, in the case of quantum mechanics, surprising).

This book can be used to supplement any general chemistry textbook. It lets the instructor choose whichever general chemistry book covers basic concepts and descriptive chemistry in a way which seems most appropriate for the students. Of course descriptive chemistry is an essential component of every freshman course. My own class includes demonstrations in every lecture and coverage of a very wide range of chemical applications. The challenge to us was to keep the strong coverage of descriptive chemistry characteristic of the best modern texts, yet elevate the mathematical level to something more appropriate for our better students. Many important aspects of chemistry can only be memorized, not understood, without appeal to mathematics. For example, the basic principles behind classical physics are quite familiar to most of these students. Almost all of them have used $\vec{F} = m\vec{a}$, potential energy, and Coulomb's law; many molecular properties follow simply from an understanding of how charges interact. When these students move on to study organic reaction mechanisms or protein folding, much of their comprehension will depend on how well they understand these basic concepts.

Later I use the same principles to show something is wrong with any classical interpretation of atomic and molecular structure. Quantum mechanics allows us to predict the structure of atoms and molecules in a manner which agrees extremely well with experimental evidence, but the intrinsic logic cannot be understood without equations.

0.1.1 Structure of the Second Edition

In this new edition, I have tried to integrate chemical applications systematically throughout the book. Chapter 1 reviews the pre-calculus mathematical concepts every general chemistry student will need by the end of a college chemistry class. The level of coverage in this chapter is essentially the same as the SAT II Math Level 1C exam. However, this chapter should not be skipped even by mathematically advanced students—it emphasizes the connections from algebra and trigonometry to chemical concepts such as solubility products, balancing equations, and half-lives. It also establishes the notational conventions.

Chapter 2 presents the basics of differential and integral calculus. I use derivatives of one variable extensively in the rest of the book. I also use the *concept* of integration as a way to determine the area under a curve, but the students are only asked to gain a qualitative understanding (at a level which allows them to look up integrals in a table), particularly in the first five chapters. Multivariate calculus is never used.

Chapter 3 is the physics chapter. The first edition jumped into Newton's laws written with calculus ($\vec{F} = d\vec{p}/dt$), which many students found overwhelming. *This version moderates that introduction by presenting the concepts of force and energy more gradually. New to this edition is an extensive discussion of atoms and molecules as charged objects with forces and potential energy (this discussion was previously much later in the book).*

Chapter 4 is an introduction to statistics (the Gaussian and Boltzmann distributions), and includes a wide range of applications (diffusion, error bars, gas kinetic energy, reaction rates, relation between equilibrium constants and energy changes). It is in my opinion a very important chapter, because it provides a quantitative foundation for the most important equations they will see in their general chemistry textbook. It also attempts to address the continuing problem of statistical illiteracy. Our students will spend the rest of their lives hearing people lie to them with statistics; I want to start to give them the tools to separate fact from fiction.

Chapter 5 takes the student through fundamental quantum mechanics. The perspective is quite different than what is found in most texts; I *want* students to be surprised by the results of quantum mechanics, and to think at least a little about the philosophical consequences. *This edition has a much longer discussion of chemical applications (such as NMR/MRI).*

I believe essentially all of the material in the first five chapters is accessible to the advanced general chemistry students at most universities. The final three chapters are written at a somewhat higher level on the whole. Chapter 6 introduces Schrödinger's equation and rationalizes more advanced concepts, such as hybridization, molecular orbitals, and multielectron atoms. It does the one-dimensional particle-in-a-box very thoroughly (including, for example, calculating momentum and discussing nonstationary states) in order to develop qualitative principles for more complex problems.

Chapter 7 covers the kinetic theory of gases. Diffusion and the one-dimensional velocity distribution were moved to Chapter 4; the ideal gas law is used throughout the book. This chapter covers more complex material. *I have placed this material later in this edition, because any reasonable derivation of $PV = nRT$ or the three-dimensional speed distribution really requires the students to understand a good deal of freshman physics. There is also significant coverage of "dimensional analysis": determining the correct functional form for the diffusion constant, for example.*

Chapter 8 (which can be covered before Chapter 7 if desired) provides a very broad overview of molecular spectroscopy and the origins of color. The topics range all the way from rainbows and peacock feathers to microwave ovens and the greenhouse effect. Once again, the emphasis is on obtaining an understanding of how we know what we know about molecules, with mathematics kept to a minimum in most sections.

0.1.2 Additional Resources

This edition features a vastly increased number of end-of-chapter problems, and answers for about half of those problems at the end of the text. It also has supplementary material available in two different forms:

1. *The Physical Basis of Chemistry Web Site* is accessible from the Harcourt/Academic Press Web site:

    `http://www.harcourt-ap.com`,

    and also from a Web site at Princeton University:

    `http://www.princeton.edu/~wwarren`.

    It contains images and QuickTime movies geared to each of the individual chapters. Much of this material was originally created by Professor Kent Wilson and the Senses Bureau at University of California, San Diego (although in many cases the slides have been adapted to match notation in the text), and all of the material may be freely used for noncommercial purposes with acknowledgment. The Web site will also contain additional problems, the inevitable typographical corrections, and links to other useful chemistry sources.

2. From time to time, a verbatim copy of the Web site will be written to compact disk, and copies made available at no charge to adopting instructors by writing to Harcourt/Academic Press, 525 B St., Suite 1900, San Diego, CA 92101 or by arrangement with your Harcourt sales representative.

I am grateful to many of my colleagues and former students for excellent suggestions. As with the first edition, I hope that the students learn even half as much by using this book as I did by writing it.

Warren S. Warren
Princeton, New Jersey
May, 1999

0.2 Preface to the First Edition

Every attempt to employ mathematical methods in the study of chemical questions must be considered profoundly irrational and contrary to the spirit of chemistry. If mathematical analysis should ever hold a prominent place in chemistry—an aberration which is happily almost impossible—it would occasion a rapid and widespread degeneration of that science.

Auguste Comte (1798–1857)
in Philosophie Positive (1830)

I am convinced that the future progress of chemistry as an exact science depends very much upon the alliance with mathematics.

A. Frankland (1825–1899)
in Amer. J. Math. $\underline{1}$, 349 (1878)

Frankland was correct.

This book is mainly intended as a supplement for the mathematically sophisticated topics in an advanced freshman chemistry course. My intent is not to force-feed math and physics into the chemistry curriculum. It is to reintroduce just enough to make important results *understandable* (or, in the case of quantum mechanics, surprising). We have tried to produce a high-quality yet affordable volume, which can be used in conjunction with any general chemistry book. This lets the instructor choose whichever general chemistry book covers basic concepts and descriptive chemistry in a way which seems most appropriate for the students. The book might also be used for the introductory portions of a junior-level course for students who have not taken multivariate calculus, or who do not need the level of rigor associated with the common one-year junior level physical chemistry sequence; for example, an introduction to biophysical chemistry or materials science should build on a foundation which is essentially at this level.

The book grew out of supplementary lecture notes from the five years I taught advanced general chemistry at Princeton University. Placement into this course is based almost exclusively on math SAT scores—no prior knowledge of chemistry is assumed. Most of the students become science or engineering majors, and they have a broad range of interests, but the strongest common denominator is interest in and aptitude for mathematics.

Picking a text book for this group of students proved to be a difficult problem. The most important change in freshman chemistry books over the last decade has been the introduction of color to illustrate descriptive chemistry. The importance of this advance should not be minimized—it helps bring out the elegance that exists in the practical aspects of chemistry. However, it has dramatically increased the cost of producing textbooks, and as a result it has become important to "pitch" these books to the widest possible audience. In general that has meant a reduction in the level of mathematics. Most modern textbooks mainly differ in the order of presentation of the material and the style of the chapters on descriptive chemistry—and almost all of them omit topics which require a little more mathematical sophistication. Thus the challenge to us was to keep the strong coverage of descriptive chemistry characteristic of the best modern texts, yet elevate the mathematical level to something more appropriate for our better students.

In fact, many important aspects of chemistry can only be memorized, not understood, without appeal to mathematics. For example:

The basic principles behind classical mechanics are quite familiar to most of these students. Almost all of them have used $F = m\vec{a}$, or can understand that a charge going around in a circle is a current. It is easy to use only these concepts to prove that something is wrong with any classical interpretation of atomic and molecular structure. Quantum mechanics allows us to predict the structure of atoms and molecules in a manner which agrees extremely well with experimental evidence, but the intrinsic logic cannot be understood without equations.

The structure of molecules is generally explained by concepts which are simple and usually correct (for example, VSEPR theory), but clearly based on some very stringent assumptions. However, the test of these theories is their agreement with experiment. It is important to understand that modern spectroscopic techniques let us measure the structures of molecules to very high precision, but only because the experimental data can be given a theoretical foundation.

Statistics play a central role in chemistry, because we essentially never see one molecule decompose, or *two* molecules collide. When 1 g of hydrogen gas burns in oxygen to give water, 6×10^{23} hydrogen atoms undergo a fundamental change in their energy and electronic structure! The properties of the reactive mixture can only be understood in terms of averages. There is no such thing as the pressure, entropy or temperature of a single helium atom—yet temperature, entropy and pressure are macroscopic, measurable, averaged quantities of great importance.

The concepts of equilibrium as the most probable state of a very large system, the size of fluctuations about that most probable state, and entropy (randomness) as a driving force in chemical reactions, are very useful and not that difficult. We develop the Boltzmann distribution and use this concept in a variety of applications.

In all cases, I assume that the students have a standard general chemistry book at their disposal. Color pictures of exploding chemical reactions (or for that matter, of hydrogen atom line spectra and lasers) are nice, but they are already contained in all of the standard books. Thus color is not used here. The background needed for this book is a "lowest common denominator" for the standard general chemistry books; in addition, I assume that students using this book are at least taking the first semester of calculus concurrently.

I wish to thank the students who have used previous versions of this book, and have often been diligent in finding errors; and Randy Bloodsworth, who found still more of the errors I missed. Useful suggestions have come from a variety of experienced instructors over the last few years, most notably Professor Walter Kauzmann and the late Miles Pickering, Director of Undergraduate Laboratories at Princeton.

Any suggestions or corrections would be appreciated. I hope that the students learn even half as much by using this book as I did by writing it.

Warren S. Warren
Princeton, New Jersey
May 1993

Chapter 1

The Tools of the Trade: Mathematical Concepts

No human investigation can be called real science if it cannot be demonstrated mathematically.

Leonardo da Vinci (1452–1519)

This chapter will review the fundamental mathematical concepts (algebra and trigonometry) needed for a quantitative understanding of college-level chemistry and physics. Virtually all of this material is covered in high-school mathematics classes, but often the connection to real scientific applications is not obvious in those classes. In contrast, the examples used here will frequently involve chemical and physical concepts that will be covered in detail in later chapters or in the later parts of a standard freshman chemistry book. Here they will be treated as math problems; later you will see the underlying chemistry.

1.1 UNITS OF MEASUREMENT

Chemistry and physics are experimental sciences, based on measurements. Our characterization of molecules (and of everything else in the universe) rests on observable physical quantities, expressed in units that ideally would be precise, convenient and reproducible. These three requirements have always produced trade-offs. For example, the English unit of length *inch* was defined to be the length of three barleycorns laid end to end—a convenient and somewhat reproducible standard for an agricultural society. When the metric system was developed in the 1790s, the meter was defined to be

1/10,000,000 of the best current estimate of distance from the equator to the North Pole along the Prime Meridian. Unfortunately, this definition was not convenient for calibrating meter sticks. The *practical* definition was based on the distance between two scratches on a platinum-iridium bar. This bar was termed the *primary standard*. Copies (secondary standards) were calibrated against the original and then taken to other laboratories.

The most important modern system of units is the SI system, which is based around seven primary units: time (second, abbreviated s), length (meter, m), temperature (Kelvin, K), mass (kilogram, kg), amount of substance (mole, mol), current (Amperes, A) and *luminous intensity* (candela, cd). The candela is mainly important for characterizing radiation sources such as light bulbs. Physical artifacts such as the platinum-iridium bar mentioned above no longer define most of the primary units. Instead, most of the definitions rely on fundamental physical properties, which are more readily reproduced. For example, the second is defined in terms of the frequency of microwave radiation that causes atoms of the isotope cesium-133 to absorb energy. This frequency is defined to be 9,192,631,770 cycles per second (Hertz) —in other words, an instrument which counts 9,192,631,770 cycles of this wave will have measured exactly one second. Commercially available *cesium clocks* use this principle, and are accurate to a few parts in 10^{14}.

The meter is defined to be the distance light travels in a vacuum during 1/299,793,238 of a second. Thus the speed of light c is *exactly* 299,793,238 meters per second. The units are abbreviated as $m \cdot s^{-1}$ (the "·" separates the different units, which are all expressed with positive or negative exponents) or as m/s. More accurate measurements in the future will sharpen the definition of the meter, not change this numerical value. Similarly, the unit of temperature above absolute zero (Kelvin) is defined by setting the "triple point" of pure water (the only temperature where ice, water, and water vapor all exist at equilibrium) as 273.16K. These inconvenient numerical values were chosen instead of (for example) exactly 3×10^8 m $\cdot$ s^{-1} or 273K because the meter, the second and degree Kelvin all predated the modern definitions. The values were calculated to allow improved accuracy while remaining as consistent as possible with previous measurements.

The definition of the kilogram is still based on the mass of a standard metal weight kept under vacuum in Paris. The mole is defined to be the number of atoms in exactly .012 kg of a sample containing only the most common isotope of carbon (carbon-12). This means that determining Avogadro's number (the number of atoms in a mole) requires some method for counting the atoms in such a sample, or in another sample which can be related to carbon. The most accurate modern method is to determine the spacing between atoms in a single crystal of silicon (Problem 1.1). Silicon is used instead of carbon because it is far easier to produce with extremely high purity. This spacing must be combined with measurements of the density and of the relative atomic weights of carbon and silicon (including the mixture of different naturally occurring isotopes of each one) to determine Avogadro's number (6.0221367×10^{23} mol^{-1}). Each of these measurements has its own sources of uncertainty, which all contribute to the finite accuracy of the final result.

In principle, Avogadro's number could be used to eliminate the standard kilogram mass. We could define Avogadro's number to be exactly 6.0221367×10^{23}, then define .012 kg as the mass of one mole of carbon-12. However, we can determine the mass of a metal weight with more accuracy than we can count this large number of atoms.

All other physical quantities have units that are combinations of the primary units. Some of these *secondary units* have names of their own. The most important of these for our purposes are listed in Table 1.1.

TABLE 1.1 ▶ Common SI Secondary Quantities and their Units

Secondary Quantity	Abbreviation	Unit	Equivalent in Other Units
Charge	q	Coulomb (C)	$A \cdot s$
Energy	$E; U; K$	Joule (J)	$kg \cdot m^2 \cdot s^{-2}$
Frequency	ν	Hertz (Hz)	s^{-1}
Force	F	Newton (N)	$kg \cdot m \cdot s^{-2}$
Pressure	P	Pascal (Pa)	$kg \cdot m^{-1} \cdot s^{-2}$ (force per unit area)
Power or intensity	I	Watt (W)	$kg \cdot m^2 \cdot s^{-3}$ (energy per second)
Volume	V	—	m^3

Because this book covers a wide rage of subfields in chemistry and physics, we will use many different abbreviations. To avoid confusion, notice that in Table 1.1 (and throughout this book) units are always written with normal (Roman) type. Variables or physical quantities are always either Greek characters or written in *italic type*. Thus, for example, "m" is the abbreviation for meters, but "m" is the abbrevation for mass.

The kilogram, not the gram, is the primary unit of mass in the SI system, so care must be taken to use the correct units in formulas. For example, the formula for kinetic energy K is $K = ms^2/2$. If m is the mass in kg and s is the speed in $m \cdot s^{-1}$, K is in Joules. The kinetic energy of a 1 g mass moving at $1\ m \cdot s^{-1}$ is .0005 J, not 0.5 J.

Prefixes can be used with all of the primary and secondary units to change their values by powers of ten (Table 1.2). Note the abbreviations for the units. Capitalization is important; meters and moles per liter (molar), or mill- and mega-, differ only by capitalization. There are prefixes for some of the intermediate values (for example, centi- is 10^{-2}) but the common convention is to prefer these prefixes, and write 10 mm or .01 m instead of 1 cm.

Since the primary unit of length is the meter, the secondary unit of volume is the cubic meter. In practice, though, the chemical community measures volume in liters and concentration in moles per liter, and often measures temperature in degrees Celsius (labeled °C, not C, to avoid confusion with the abbreviation for charge). Other

TABLE 1.2 ▶ Common Prefixes

Prefix	Example	Numerical Factor
femto-	femtosecond (fs)	10^{-15}
pico-	picomole (pmol)	10^{-12}
nano-	nanometer (nm)	10^{-9}
micro-	micromolar (μM)	10^{-6}
milli-	milliliter (mL)	10^{-3}
—		10^{0}
kilo-	kilogram (kg)	10^{3}
mega-	megapascal (MPa)	10^{6}
giga-	gigawatt (GW)	10^{9}
tera-	terahertz (THz)	10^{12}

non-SI units in common use are listed in Table 1.3 below. For instance, the ideal gas law $PV = nRT$ in SI units uses pressure in Pascals, volume in cubic meters, and temperature in Kelvin. In that case the ideal gas constant R $= 8.3143510$ J $\cdot$ K^{-1} $\cdot$ mol^{-1}. However, it is also quite common to express pressures in atmospheres or torr. One torr is the pressure exerted by a 1 mm mercury column at the Earth's surface (the area of the column does not matter), and 1 atm is the same as the pressure exerted by a 760 mm mercury column. These alternative units require different values of R (for example, R $= 0.08206 \cdot$ L $\cdot$ atm $\cdot$ K^{-1} $\cdot$ mol^{-1}).

TABLE 1.3 ▶ Common non-SI Units and their SI Equivalents

Non-SI Quantity	Unit	Equivalent in other units
Concentration	molar (M)	mol $\cdot$ L^{-1} (volume of total solution)
	molar (m)	mol $\cdot$ kg^{-1} (mass of solvent)
Energy	kilojoule per mole (kJ $\cdot$ mol^{-1})	1.660540×10^{-21} J
	electron volt (eV)	1.602177×10^{-19} J or 96.4753 kJ $\cdot$ mol^{-1}
	calorie (cal)	4.184 J
Length	Angstrom (Å)	10^{-10} m $= 100$ pm
Mass	atomic mass unit (amu) or Dalton (Da)	$1.6605402 \times 10^{-27}$ kg
Pressure	atmosphere (atm)	$101{,}325$ Pa
	bar	$100{,}000$ Pa
	torr	$1/760$ atm; 133.32 Pa
Temperature	degrees Celsius (°C)	°C $=$ K $- 273.15$
Volume	liter (L)	$.001$ m^3

The energies associated with chemical processes are inconveniently small when expressed in Joules. For example, the dissociation energy for the bond in the H_2 molecule is 7.17×10^{-19} J. It is thus more common to write the energy associated with breaking one mole of such bonds (432 kJ $\cdot$ mol^{-1}). Another convenient energy unit is the electron volt (eV), which (as the name implies) is the energy picked up by an electron when it is moved across a potential of one volt. We will discuss this more in Chapters 3 and 5.

The atomic mass unit (amu) is 1/12 of the mass of a single atom of carbon-12, and as the name implies, is the usual unit for atomic masses. It is also commonly called the *Dalton* (abbreviated Da) in biochemistry books, and is equivalent to 1 g$\cdot$ mol^{-1}. The mass of a single atom of the most common isotope of hydrogen (one proton and one electron) is 1.007825 amu. Naturally occurring hydrogen also contains a second isotope: about 0.015% of the atoms have one neutron and one proton, and this isotope (called *deuterium*, abbreviated D) has mass 2.0141018 amu. This makes the average mass of naturally occurring hydrogen (the mass of one mole divided by Avogadro's number) about equal to 1.00797 amu:

$$\text{Avg. mass} = (0.99985) \cdot 1.007825 + (0.00015 \cdot 2.0141018) = 1.00797 \text{ amu}$$

However, the fraction of deuterium can vary in naturally occurring samples, because isotopic substitution can slightly change chemical properties. Normal water (H_2O) boils at $100°C$ (at 1 atm pressure) and freezes at $0°C$; heavy water (D_2O) boils at $101.42°C$ and freezes at $3.82°C$.

The task of reconciling experimental measurements in many different laboratories to produce the best possible set of fundamental physical constants is assigned to CODATA (the Committee on Data for Science and Technology), established in 1966 by the International Council of Scientific Unions. Roughly every ten years this group releases a new set of constants. Appendix A presents the 1998 values. Each value also has associated *error bars*, which we will explain in more detail in Chapter 4.

1.2 COMMON FUNCTIONS AND CHEMICAL APPLICATIONS

1.2.1 Definition of Functions and Inverse Functions

When we have a relation between two variables such as $y = x^2$, we say that y is a *function* of x if there is a unique value of y for each value of x. Sometimes we also write $y = f(x)$ to emphasize the function itself (in this case, the function f corresponds to the operation of squaring).

Every function has a *domain* (the set of all permitted values of x) and a *range* (the set of all permitted values of y). For this equation, we can pick any real value of x and produce a value for y, so the domain is $[-\infty, \infty]$. However, not all values of y are possible. This particular function has a minimum at ($x = 0$, $y = 0$) hence the range is $[0, \infty]$.

A function must have a unique value of y for each value of x, but it need not have a unique value of x for each value of y. If it does have a unique value of x for each value of y, we can also define an ***inverse function***, $x = g(y)$. We could rearrange $y = x^2$ and write $x = \sqrt{y}$ (taking the square root), but there are two values of x (plus and minus) associated with a single positive value of y, and no real values of x associated with a negative value of y. So the square root is only a function if the domain and range are restricted to nonnegative numbers. With that restriction, squaring and taking square roots are inverse functions of one another—if you do them in succession, you get back to your original number.

1.2.2 Polynomial Functions

Polynomials are among the simplest functions we will use. In general, a polynomial has the form:

$$y = a_0 + a_1 x + a_2 x^2 + a_3 x^3 + \cdots + a_n x^n = \sum_{i=0}^{n} a_i x^i \qquad (1.1)$$

where the coefficients $a_0, a_1, a_2 \ldots$ are numbers which do not depend on x. Equation 1.1 uses a common shorthand (summations represented by the Greek character Σ) to reduce an expression with many terms into a simpler schematic form. The lower limit of the summation is presented under the summation character; the upper limit is above the character.

Notice that only integral powers of x appear in the expression. Polynomial equations where the highest power of x (the ***order***) is two, such as $y = 2x^2 + 6x + 1$ or $y = ax^2 + bx + c$, are also called ***quadratic equations***. Third-order polynomials are called ***cubic equations***.

Quadratic equations arise frequently in the mathematical descriptions of common physical and chemical processes. For instance, silver chloride is only very slightly soluble in water. It has been determined experimentally that the ***solubility product*** K_{sp} of silver chloride at 25C is 1.56×10^{-12} M^2, meaning that in a saturated solution concentrations of silver ion and chloride ion satisfy the relationship

$$K_{sp} = [Ag^+][Cl^-] = 1.56 \times 10^{-10} \; M^2 \qquad (1.2)$$

Following the usual convention, we will now express all concentrations in moles per liter, and drop the units "M^2" from the solubility product expression. If solid silver chloride is added to water, dissociation of the solid will create an equal number of silver ions and chloride ions, hence the concentrations of silver ion and chloride ion will be the same. If we substitute $x = [Ag^+] = [Cl^-]$ into Equation 1.2, this implies

$$x^2 = 1.56 \times 10^{-10}$$
$$x = [Ag^+] = [Cl^-] = 1.25 \times 10^{-5} \text{ moles per liter} \qquad (1.3)$$

Equation 1.2 also has a negative solution, but we cannot have a negative amount of dissociated silver chloride, hence the domain of x is the set of nonnegative numbers. Thus one liter of saturated silver chloride solution contains 1.25×10^{-5} moles of dissolved silver chloride.

On the other hand, suppose we add silver chloride to a .01 M sodium chloride solution. Now if the number of moles dissolved in one liter is x, we have

$$
\begin{aligned}
[Ag^+] &= x; [Cl^-] = x + .01 \\
(x)(x + .01) &= 1.56 \times 10^{-10} \\
x^2 + .01x - (1.56 \times 10^{-10}) &= 0
\end{aligned}
\tag{1.4}
$$

In high school algebra you frequently used factoring to solve quadratic equations. However, in this and virtually all other real problems you will encounter in chemistry, neither the numerical coefficients nor the solutions will turn out to be integers, so factoring is not useful. The general solutions to the equation $ax^2 + bx + c = 0$ are given by the expression

$$
\text{If } ax^2 + bx + c = 0, x = \frac{-b \pm \sqrt{b^2 - 4ac}}{2a}
\tag{1.5}
$$

Here $a = 1$, $b = .01$, and $c = 1.56 \times 10^{-10}$, so Equation 1.5 gives $x = 1.56 \times 10^{-8}$ (and a negative solution, which again is not physically reasonable.) Note that far less silver chloride can be dissolved in a sodium chloride solution than in pure water—this is called the ***common ion effect***.

Often it is possible to use approximate methods to avoid the tedium of solving Equation 1.5. In this problem, the small value of the solubility product tells us that the final concentration of the chloride ions will not be affected much by the dissolved silver chloride. So if we write $[Cl^-] = 0.01 + x \approx 0.01$, Equation 1.4 reduces to $0.01x = 1.56 \times 10^{-10}$, which leads to the same final answer (to the number of significant digits used here). It is often a good strategy to try an approximation, obtain a (tentative) answer, then plug the answer back into the original equation to verify its accuracy.

Cubic and higher order polynomial expressions also arise naturally in a wide range of problems in chemistry, particularly in solubility and equilibrium problems. If we try to dissolve lead (II) chloride ($PbCl_2$) instead of silver chloride, the solubility product expression becomes

$$
K_{sp} = [Pb^{2+}][Cl^-]^2 = 1.6 \times 10^{-5}
\tag{1.6}
$$

Each formula unit of $PbCl_2$ dissolved in water creates *two* units of chloride, and one of lead. So in one liter of water, x moles of dissolved $PbCl_2$ gives

$$
[Pb^{2+}] = x; [Cl^-] = 2x; 4x^3 = 1.6 \times 10^{-5}; x = 1.6 \times 10^{-2} \text{ moles per liter}
\tag{1.7}
$$

Suppose we try to dissolve lead (II) chloride in a 0.1 M solution of sodium chloride. Now x moles of dissolved lead (II) chloride give concentrations of

$$[Pb^{2+}] = x;$$
$$[Cl^-] = 2x + 0.1; \tag{1.8}$$
$$x(2x + 0.1)^2 = 4x^3 + 0.4x^2 + 0.01x = 1.6 \times 10^{-5}$$

The solution for a general cubic equation (analogous to Equation 1.5) is vastly more tedious, and never used in practice. This problem can be solved by assuming that the small amount of dissolved $PbCl_2$ does not substantially change the chloride concentration ($x \ll 0.1$), which gives

$$[Pb^{2+}] = x; [Cl^-] = 2x + 0.1 \approx 0.1;$$
$$0.01x = 1.6 \times 10^{-5}; x = 1.6 \times 10^{-3} \text{ moles per liter} \tag{1.9}$$

You can plug $x = 1.6 \times 10^{-3}$ into Equation 1.8 to verify that the approximation is reasonable. If the concentration of chloride were lower (say 10^{-2} M) the approximation method would not work very well, and we would have to solve the equation

$$[Pb^{2+}] = x; [Cl^-] = 2x + 0.01;$$
$$x(2x + 0.01)^2 = 4x^3 + 0.04x^2 + 0.0001x = 1.6 \times 10^{-5} \tag{1.10}$$

In the days before the common availability of personal computers, one then had to resort to successive approximations. Today such a problem would be solved graphically, numerically on a graphing calculator, or on a computer; the solution is $x = 0.013$ moles per liter.

1.2.3 Trigonometric Functions

Another important class of functions encountered in chemistry and physics is the *trigonometric functions*. Consider the equation $x^2 + y^2 = 1$. The set of all points in a plane that satisfy this equation is a circle with radius 1 (Figure 1.1). Any position on the circle could be labeled by the length θ of the arc which stretches counterclockwise from the positive x-axis to that point. Since the circle has circumference 2π, only values of θ between 0 and 2π are needed to describe the whole circle.

We can give the same label θ to the angle that creates this arc. In this case, we refer to the angle in units of *radians*, and thus 2π corresponds to a complete circle. Radians might seem superficially to be an inconvenient unit for measuring angles. In fact, they turn out to be the most natural unit, as we will see when we discuss *derivatives* in the next chapter.

We can also use Figure 1.1 to define trigonometric functions. The x-coordinate gives the cosine of the angle θ, written $\cos \theta$. The y-coordinate gives the sine of the angle θ, written $\sin \theta$. Figure 1.1 shows that $\sin^2 \theta + \cos^2 \theta = 1$. We also define the tangent of the angle θ, written $\tan \theta$, as the ratio $\sin \theta / \cos \theta$.

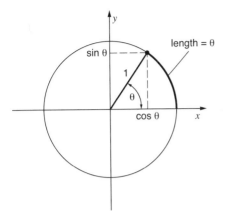

FIGURE 1.1 ▶ Definition of the sine and cosine function, in terms of positions on a circle with radius 1.

Now suppose we move counter-clockwise along the circle at a constant speed, which we will call ω. ω has units of radians per second, and is also called the ***angular velocity***. The x- and y-coordinates will vary with time as shown in Figure 1.2. Notice that the waveform is the same for the cosine (x-coordinate) and the sine (y-coordinate) except for a shift of one-quarter cycle. The ***frequency*** of the sine wave, commonly de-noted by the symbol ν, is the number of cycles per second. This unit is given the special name of Hertz. Since there are 2π radians in one cycle, $\omega = 2\pi\nu$. One complete cycle requires a time $T = 2\pi/\omega = 1/\nu$, which we call the ***period*** of the sine wave (seconds per cycle).

Don't let the definition based on tracking around a circle fool you—sine and co-sine waves appear in many problems in chemistry and physics. The motion of a mass

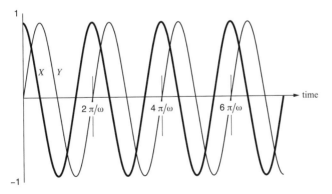

FIGURE 1.2 ▶ Sine (y) and cosine (x) components of motion at a constant angular velocity ω along a circular path.

suspended from a spring, or of a pendulum with small swings, is sinusoidal. In addition, what we call *light* is a combination of an electric field and a magnetic field, as discussed in Chapter 3. If these fields are sine waves at the same frequency (such as 5×10^{14} Hertz), the eye perceives a well-defined color (in that case, red). One of the results of quantum mechanics (Chapter 5) will be that such a single frequency *electromagnetic* wave consists of particles (called *photons*) with a well-defined energy. The cesium clock mentioned in Section 1.1 absorbs photons, and in the process an electron moves into a higher energy level.

Each of the trigonometric functions also has an inverse function. For example, as θ goes from 0 to π in Figure 1.1, $x = \cos\theta$ goes from $+1$ to -1. No two values of θ in this domain give the same value of x. Therefore we can define an inverse cosine function, $\theta = \arccos x$, which gives a single value between 0 and π for each value of x between 1 and -1. For example, the only value of θ between 0 and π which gives $\cos\theta = 0$ is $\theta = \pi/2$, so $\arccos(0) = \pi/2$. Some books refer to $\arccos\theta$ as "$\cos^{-1}\theta$," but that notation is confusing because the inverse function is *not* the same as the reciprocal; $1/\cos(0) = 1$, not $\pi/2$.

1.3   VECTORS AND DIRECTIONS

Force, momentum, velocity and acceleration are examples of *vector quantities* (they have a direction and a magnitude) and are written in this book with an arrow over them. Other physical quantities (for example, mass and energy) which do not have a direction will be written without an arrow. The directional nature of vector quantities is often quite important. Two cars moving with the same velocity will *never* collide, but two cars with the same speed (going in different directions) certainly might!

In general, three coordinates are required to specify the magnitude and direction of a vector in three-dimensional space. The most common system for specifying these parameters is *Cartesian coordinates*, which specify the projections (x, y, z) of the vector along three mutually perpendicular axes (Figure 1.3, left). Sometimes we will refer only to the component of a vector quantity along a specific direction, which we will signify by a subscript and without boldface. Thus the velocity vector $\vec{v}$ has components (v_x, v_y, v_z) along the three Cartesian coordinates. The magnitude $|\vec{v}|$ of the velocity vector (which we will usually call the *speed s*) is given by

$$s = |\vec{v}| = \sqrt{v_x^2 + v_y^2 + v_z^2} \tag{1.11}$$

It is also sometimes convenient to specify only the direction of a vector. This is done by introducing *unit vectors*, which are defined to have length one. Unit vectors are signified by a caret (ˆ) instead of an arrow. Thus $\vec{v} = |\vec{v}|\,\hat{v} = s\hat{v}$.

We will make one exception to these rules for simplicity: we will write the components of the position vector $\vec{r}$ as simply (x, y, z), and its magnitude as r. Hence $r = \sqrt{x^2 + y^2 + z^2}$ and $\vec{r} = r\hat{r}$.

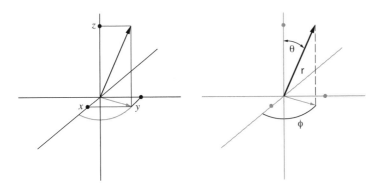

FIGURE 1.3 ▶ The position of a vector can be expressed in Cartesian coordinates (x, y, z), left, or spherical coordinates (r, θ, ϕ), right.

Cartesian coordinates have some advantages for describing vectors. For instance, we can add two vectors by adding the individual components

$$(2, 3, 5) + (1, 1, 6) = (3, 4, 11).$$

Despite these advantages, other coordinate systems often turn out to be more useful than Cartesian coordinates. For example, the interaction energy between a proton and an electron depends only on distance between them, not on the direction; we say that the potential generated by the proton is **spherically symmetric**. As a result, when we discuss the possible energy levels for the electron in a hydrogen atom, the expressions will be far simpler in **spherical coordinates**, which specify the length of the vector (r) and two angles to give the orientation (Figure 1.3, right). The angle θ is the angle between the vector and the z-axis. The angle ϕ is the angle that the projection of the vector down into the xy-plane makes with the x-axis.

Every point (x, y, z) in Cartesian coordinates corresponds to a unique value of (r, θ, ϕ), with $r > 0$, $0 \leq \theta \leq \pi$, and $0 \leq \phi < 2\pi$, except for points along the z-axis (where ϕ is undefined). Values of ϕ outside of this range can be moved into the range by adding some multiple of 2π; for example, $\phi = -\pi/2$ is the same as $\phi = 3\pi/2$. We can convert between spherical and Cartesian coordinates by the following relationships:

$$
\begin{aligned}
x &= r \sin\theta \cos\phi \\
y &= r \sin\theta \sin\phi \\
z &= r \cos\theta \\
r &= \sqrt{x^2 + y^2 + z^2}; \\
\theta &= \arccos\left(z/\sqrt{x^2 + y^2 + z^2}\right); \\
\phi &= \begin{cases} \arctan(y/x) & \text{if } x > 0 \\ \pi + \arctan(y/x) & \text{if } x < 0 \end{cases}
\end{aligned}
\tag{1.12}
$$

For example, $(x, y, z) = (1, 1, 0)$ is the same as $(r, \theta, \phi) = \left(\sqrt{2}, \pi/2, \pi/4\right)$; $(x, y, z) = (0, 2, 0)$ is the same as $(r, \theta, \phi) = (2, \pi/2, \pi/2)$.

Spherical coordinates have the advantage that the length is immediately obvious (it is the r coordinate), but they have some disadvantages as well. For example, vectors in spherical coordinates cannot be added just by adding their components.

1.4   EXPONENTIALS AND LOGARITHMS

Exponentials and logarithms appear in many formulas in chemistry. We have already encountered them in the definitions of prefixes in Table 1.2, which are essentially a shorthand to avoid large powers of ten (we can write 17 ps instead of 1.7×10^{-11} s). In addition to powers of 10, we frequently use powers of $e = 2.7183\ldots$ and occasionally use powers of 2. The number e (base of natural logarithms) arises naturally in calculus, for reasons we will discuss briefly later (calculus classes explain it in great detail). Powers of e occur so often that a common notation is to write $\exp(x)$ instead of e^x.

Powers of two arise naturally in digital electronics. Bits are stored in a computer as 1 or 0. Each letter on the computer keyboard is stored as a unique combination of eight bits, called a byte. There are $2^8 = 256$ possible combinations. Bytes are in turn addressed on computer chips in powers of two. A small personal computer might have "64 MB RAM," or 64 megabytes of random access memory, but the prefix "mega-" is deceptive (and international scientific organizations have proposed a replacement). In computer language, because of the internal construction of integrated circuits, it means $2^{20} = 1,048,576$. In scientific notation (and in everything we do in chemistry), the prefix "mega" means *exactly* 10^6.

1.4.1 Properties of Exponentials

Most of the properties of exponentials are the same for 10^x, 2^x, or e^x. For example, $10^1 \cdot 10^2 (= 10 \cdot 100) = 100^{1+2} (= 1000)$, and $(10^2)^2 = 10^{2\cdot2} = 10^4$. In general we can write:

$$
\begin{aligned}
2^{a+b} &= 2^a \cdot 2^b \\
10^{a+b} &= 10^a \cdot 10^b \\
e^{(a+b)} &= e^a \cdot e^b
\end{aligned}
\tag{1.13}
$$

$$
\begin{aligned}
2^{ab} &= \left(2^a\right)^b = \left(2^b\right)^a \\
10^{ab} &= \left(10^a\right)^b = \left(10^b\right)^a \\
e^{ab} &= \left(e^a\right)^b = \left(e^b\right)^a
\end{aligned}
\tag{1.14}
$$

with similar results for any other (positive) number. The power need not be integral, but the convention is that a^b is positive if a is positive. Thus, even though both (-2) and 2 give 4 when squared, we define $4^{1/2} = 2$.

The logarithm is the inverse of the exponential operation. Thus, if $y = a^x$ we define $x \equiv \log_a y$ (read as "log base a of y"). Again the common logarithms are base 10 (often written "log") and base e (often written "ln"); base 2 also appears occasionally. From the definition,

$$\ln(e^x) = x; \, e^{\ln x} = x \, (x > 0) \tag{1.15}$$

$$\log(10^x) = x; \, 10^{\log x} = x \, (x > 0) \tag{1.16}$$

Other common properties include:

$$\begin{aligned} \log ab &= \log a + \log b; \\ \log(a/b) &= \log a - \log b; \\ \log a^n &= n \log a \end{aligned} \tag{1.17}$$

Analogous formulas apply for ln.

Conversion between different bases is sometimes necessary. This can be done as follows:

$$\begin{aligned} 10^x &= [e^{(\ln 10)}]^x \text{ (from Eq. 1.15)} \\ &= e^{x \ln 10} \approx e^{(2.3026...)x} \text{ (from Eq. 1.14)} \end{aligned} \tag{1.18}$$

Similarly

$$2^x = e^{x \ln 2} \approx e^{(.694...)x} \tag{1.19}$$

Conversion between log and ln is also simple:

$$\ln x = (\ln 10)(\log x) \approx (2.3026\dots) \log x \tag{1.20}$$

The numerical factor $\ln 10 = 2.3026\dots$ pops up in some of the equations you will see in chemistry and physics; in fact, many chemical equations you see with "log" in them will actually have "2.3026 log", essentially because the equation should really contain ln.

1.4.2 Applications of Exponentials and Logarithms
▶ *Nuclear Disintegrations and Reaction Kinetics*

Exponentials play a useful role in understanding nuclear disintegrations and half-lives. For example, ^{14}C, a radioactive isotope of carbon used for "carbon dating," has a half-life $t_{1/2} = 5730$ years before it converts into stable ^{14}N. This means that the number

N of carbon-14 atoms in a sample will satisfy the following relationships:

$$N(t_{1/2}) = N(0)/2; \quad N(t_{1/2}) = N(0)/4 : \quad N(3t_{1/2}) = N(0)/8$$
$$(1.21)$$

Half of the sample will be left after $t_{1/2} = 5730$ years; half of what is left will decay in another 5730 years (in other words, one-quarter will be left after $2t_{1/2} = 11460$ years), and so forth. More generally,

$$N(t) = N(0)2^{-t/t_{1/2}} \tag{1.22}$$

Using Equation 1.19 this can also be written as

$$N(t) = N(0)e^{-kt}; \quad k = (.694\ldots)/t_{1/2} \tag{1.23}$$

Written in this form, k is called the rate constant. Rate constants also appear in chemical kinetics. For instance, in Chapter 4 we will show that the rate of a unimolecular reaction (such as an internal rearrangement of some atoms within a single molecule) changes with temperature according to the equation

$$k = A \exp{(-E_a/k_B T)} \tag{1.24}$$

Here $k_B = 1.38 \times 10^{-23}$ J $\cdot$ K^{-1} is a numerical constant called *Boltzmann's constant*. The parameters A and E_a depend on the specific chemical reaction.

▶ Hydrogen Ion Concentrations

A very broad range of hydrogen ion concentrations is encountered in chemical reactions. A 1M solution of hydrochloric acid, a very strong acid, has a hydrogen ion concentration ($[H^+]$) of about 1M. A 1M solution of sodium hydroxide (NaOH) has a hydroxide ion concentration ($[OH^-]$) of about 1M, and since the product $[H^+][OH^-] = 10^{-14}$ M^2 in water at 25C, the hydrogen ion concentration is about 10^{-14} M.

Rather than dealing with such a wide concentration range, we usually express hydrogen ion concentration (acidity) by the pH:

$$pH = -\log{[H^+]} \tag{1.25}$$

With this definition, pH $= 0$ for a 1M HCl solution, and pH $= 14$ for a 1M NaOH solution.

The choice of pH as the usual measure of acidity is more than just a practical convenience. Voltages generated by electrochemical cells are generally proportional to the log of the concentration. Electrochemical devices that measure hydrogen ion concentration (*pH meters*) are readily available, and such a device could readily measure either the 1M or the 10^{-14} M concentrations mentioned above. This does not imply, however, that the 1M concentration can be measured to 10^{-14} M accuracy! A typical pH meter will be accurate to about .01 units over this entire range.

▶ PROBLEMS ▶ _____

In this chapter and all other chapters, answers to the starred problems can be found in the Answer Key at the end of the book.

Units of Measurement

1-1.★ Find the volume of exactly one mole of an ideal gas at "standard temperature and pressure" ($T = 273.15$K, $P = 1$ atm $= 101325$ Pa).

1-2. Einstein's famous formula $E = mc^2$, which shows that mass can be converted into energy, is written in SI units. Determine how much energy (in Joules) is created by the destruction of one gram of matter. Compare this to the energy liberated by the combustion of one gram of gasoline (50 kJ).

1-3.★ The volume per silicon atom in crystalline silicon can be measured spectroscopically. There are eight atoms in the "unit cell," which is a cube with side length 543.10196 pm. The density of crystalline silicon is 2.3291 g · mL^{-1}. The atomic weight of naturally occurring silicon is 28.086 g · mol^{-1}. Show how these numbers can be combined to give Avogadro's number.

1-4. Find the correct value for the ideal gas constant R (including the units) when pressure is expressed in Torr, volume is expressed in cubic centimeters, and temperature is expressed in degrees Kelvin.

Applications of Functions in Chemistry

1-5.★ Silver chloride is much more soluble in boiling water ($K_{sp} = 2.15 \times 10^{-8}$ at $T = 100$C) than it is at room temperature ($K_{sp} = 1.56 \times 10^{-10}$). How much silver chloride is dissolved in 1L of a saturated solution at 100C?

1-6. How much silver chloride can be dissolved in 1L of a 0.1M sodium chloride solution at 100C? Explicitly state the approximation you are using to solve this problem, and show that the approximation is valid.

1-7.★ Using a computer or graphing calculator, determine the amount of silver chloride which can be dissolved in 1L of a 10^{-4} M sodium chloride solution at 100C.

1-8. Lead iodide (PbI$_2$) dissolves in water with solubility product

$$K_{sp} = [Pb^{+2}][I^-]^2 = 1.39 \times 10^{-8}$$

at 25C. How much lead iodide is present in 1L of a saturated solution?

1-9.★ How much lead iodide can be dissolved in 1L of a 1M sodium iodide solution at 25C? Explicitly state the approximation you are using to solve this problem, and show that the approximation is valid.

1-10. Using a computer or graphing calculator, determine the amount of lead iodide which can be dissolved in 1L of a .001M sodium iodide solution.

Vectors and Directions

1-11.★ A particle is located along the x-axis in a Cartesian coordinate system, 1 unit from the origin. Find its position in spherical coordinates.

1-12. A particle at the position $(r, \theta, \phi) = (1, \pi/4, \pi/2)$. Find its position in Cartesian coordinates.

1-13. Find the geometrical object described by each of the following equations:

(a)★ $z = 2$ (b) $r = 6$

(c)★ $\phi = \pi/4$ (d) $\theta = 0$

(e)★ $\theta = \pi/4$ (f) $\theta = \pi/2$

Trigonometric Functions

1-14. Two very useful formulas for converting between sines and cosines of different angles are:

$$\sin(\theta_1 + \theta_2) = \sin\theta_1 \cos\theta_2 + \cos\theta_1 \sin\theta_2$$
$$\cos(\theta_1 + \theta_2) = \cos\theta_1 \cos\theta_2 - \sin\theta_1 \sin\theta_2$$

Use these relations and the definition of the sine and cosine (see Figure 1.1) to find the sine and cosine for $\theta = 0, \pi/4, \pi/2, 3\pi/4, \pi, 5\pi/4, 3\pi/2, 7\pi/4, 2\pi$.

1-15. Light rays bend when they pass from one substance to another (for example, from water into air; see the figure below). The equation which describes the change in direction (called *Snell's Law*) is $n_1 \sin\theta_1 = n_2 \sin\theta_2$. Here θ_1 and θ_2 are the angles the light makes with a perpendicular to the surface, as shown in the diagram. The numbers n_1 and n_2 (the *indices of refraction* of the materials) are tabulated in reference books. The values in the figure are for yellow light and room temperature.

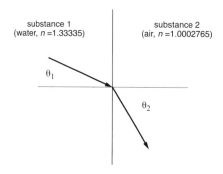

substance 1
(water, $n = 1.33335$)

substance 2
(air, $n = 1.0002765$)

θ_1

θ_2

(a)★ Find θ_2 if $\theta_1 = 45°$.

(b) Show that if θ_1 exceeds a certain value for light going from water to air (called the **critical angle** θ_c), Snell's law cannot be satisfied. In this case the light gets completely reflected instead of transmitted (this is called **total internal reflection**). Show also that Snell's law can always be satisfied for light going from air to water.

(c) Find θ_c for this system.

1-16. Over what range and domain can the arctangent function be defined if we want it to be monotonically increasing (e.g., $\arctan(\theta_2) > \arctan(\theta_1)$ if $\theta_2 > \theta_1$)?

1-17.★ Over what range and domain can the arcsine function be defined if we want it to be monotonically increasing?

Exponentials and Logarithms

1-18. Why is the restriction to $x > 0$ necessary in Equations 1.15 and 1.16?

1-19. Without using a calculator, given only that $\log 2 = 0.301$ (and of course $\log 10 = 1$), find the following logs:

(a)★ $\log 4$ (b) $\log 40$

(c)★ $\log 50$ (d) $\log 0.025$

Other Problems

1-20. At 25C, the ionization constant $K_w = [H^+][OH^-]$ for pure water is 1.00×10^{-14}; at 60C, $K_w = 9.6 \times 10^{-14}$. Find the pH of pure water at both temperatures.

1-21. Tritium (hydrogen-3) is used to enhance the explosive yield of nuclear warheads. It is manufactured in specialized nuclear reactors. The half-life of tritium is 12.3 years. If no new tritium were produced, what fraction of the world's supply of tritium would exist in 50 years?

1-22. The Shroud of Turin is a length of linen that for centuries was purported to be the burial garment of Jesus Christ. In 1988 three laboratories in different countries measured the carbon-14 content of very small pieces of the Shroud's cloth. All three laboratories concluded that the cloth had been made sometime between AD 1260 and AD 1390.

(a) If these results are valid, how much less carbon-14 is there in the Shroud than in a new piece of cloth? Use the midpoint of the age range (AD 1325) and use 5730 years for the half-life of carbon-14.

(b) If the Shroud dated from the crucifixion of Jesus (approximately AD 30), how much less carbon-14 would there be in the Shroud than in a new piece of cloth?

(c) It has very recently been proposed that the radiocarbon dating measurement might be in error because mold has grown on the Shroud over the years. Thus the measured carbon-14 content would be the average of new organic material and the old Shroud, and the Shroud would appear more recent.

Assume that the Shroud actually dates from AD 30, that the mold has the same percentage carbon content as the linen, and that the mold all grew recently (so that it looks new by carbon-14 dating). How many grams of mold would there have to be on each gram of Shroud linen to change the apparent origin date to AD 1325? (Mold was not apparent on the tested samples.)

1-23. Organic chemists use a common "rule of thumb" that the rate of a typical chemical reaction doubles as the temperature is increased by ten degrees. Assume that the constants A and E_a in Equation 1.24 do not change as the temperature changes. What must the value of E_a (called the activation energy) be for the rate to double as the temperature is raised from 25C to 35C?

1-24. Balancing chemical reactions is an application of solving multiple simultaneous linear equations. Consider, for example, the complete combustion of one mole of methane to produce carbon dioxide and water:

$$CH_4 + xO_2 \longrightarrow yCO_2 + zH_2O$$

Since atoms are not transmuted under normal chemical conditions, this can be balanced by equating the number of carbon, hydrogen, and oxygen atoms on each side

$$1 = y(\text{carbon}); \quad 4 = 2z(\text{hydrogen}); \quad 2x = 2y + z(\text{oxygen})$$

These equations can be solved by inspection: $x = 2$, $y = 1$, $z = 2$.

However, a balanced equation tells us nothing about the physical reaction pathway, or even whether or not a reaction is possible—balancing is essentially algebra. Sometimes there is not even a unique solution. To see this, balance the following equation:

$$CH_4 + xO_2 \longrightarrow wCO + yCO_2 + zH_2O$$

using $\pi/2$ moles of oxygen per mole of methane. It turns out that *any* value of x within a certain range will give a valid balanced equation (with all coefficients positive); what is this range?

Chapter 2

Essentials of Calculus
for Chemical Applications

The latest authors, like the most ancient, strove to subordinate the phenomena of nature to the laws of mathematics.

Sir Isaac Newton (1642–1727)

Perhaps the most remarkable feature of modern chemical theory is the seamless transition it makes from a microscopic level (dealing directly with the properties of atoms) to describe the structure, reactivity and energetics of molecules as complicated as proteins and enzymes. The foundations of this theoretical structure are based on physics and mathematics at a somewhat higher level than is normally found in high school. In particular, calculus provides an indispensable tool for understanding how particles move and interact, except in somewhat artificial limits (such as perfectly constant velocity or acceleration). It also provides a direct connection between some observable quantities, such as force and energy.

This chapter highlights a small part of the core material covered in a first-year calculus class (derivatives and integrals in one dimension). The treatment of integrals is particularly brief—in general we do not explicitly calculate integrals in this book. However, we will often tell you the value of some integral, and so we will very briefly summarize integration here to help you understand the concept.

2.1 DERIVATIVES

2.1.1 Definition of the Derivative

Suppose we have some "smoothly varying function" $y = f(x)$ which might look like Figure 2.1 when graphed ($y = f(x)$ on the vertical axis, x on the horizontal axis).

There is a formal definition of a "smoothly varying function", but for our purposes, what we mean is that the curve has no breaks or kinks.

We can draw a unique **tangent line** (a straight line whose slope matches the curve's slope) at each point on the curve. Recall that the slope of a line is defined as the amount y changes if x is changed by one; for example, the line $y = 3x + 6$ has a slope of three.

Three of these tangent lines are drawn on the curve in Figure 2.1. You can see qualitatively how to draw them, but you cannot tell by inspection exactly what slope to use. This slope can be found by looking at two points x_0 and $x_0 + \Delta x$, where Δx (the separation between the two points, pronounced "delta x") is small. We then determine the amount $\Delta y = f(x_0 + \Delta x) - f(x)$ that the height of the curve changes between those two points. The ratio $\Delta y / \Delta x$ approaches the slope of the tangent line, in the limit that Δx is very small, and is called the **derivative** dy/dx. Another common shorthand is to write the derivative of $f(x)$ as $f'(x)$.

The mathematical definition of the derivative is:

$$\left. \frac{dy}{dx} \right|_{x=x_0} = \lim_{\Delta x \to 0} \left(\frac{\Delta y}{\Delta x} \right) = \lim_{\Delta x \to 0} \left(\frac{f(x_0 + \Delta x) - f(x_0)}{\Delta x} \right) \tag{2.1}$$

The "d" in dy and in dx means "Δ in the limit of infinitesimally small changes." The "$|_{x=x_0}$" in Equation 2.1 just means "evaluated at the point $x = x_0$." The restriction $\Delta x \to 0$ is very important; the expression in Equation 2.1 will only give the slope of the tangent line in that limit. You can see from the illustration in Figure 2.1 that a line drawn through the two points x_0 and $(x_0 + \Delta x)$ would be close to the tangent curve, but not on top of it, because Δx is not arbitrarily small.

Much of the first semester of calculus is devoted to understanding what is meant by a "smoothly varying function," and finding the derivatives of various functions. For

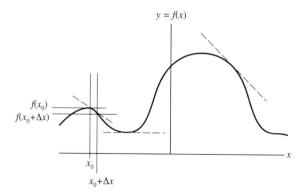

FIGURE 2.1 ▶ Graph of an arbitrary function $f(x)$. The dashed lines show tangent curves at several points. The slope of the tangent line (called the **derivative**) can be found by drawing a line between two very close points (here x_0 and $x_0 + \Delta x$).

example, suppose $y = x^3$:

$$\left.\frac{d(x^3)}{dx}\right|_{x=x_0} = \lim_{\Delta x \to 0} \left(\frac{(x_0 + \Delta x)^3 - x_0^3}{\Delta x}\right)$$

$$= \lim_{\Delta x \to 0} \left(\frac{x_0^3 + 3(\Delta x)x_0^2 + 3(\Delta x)^2 x_0 + (\Delta x)^3 - x_0^3}{\Delta x}\right)$$

$$= \lim_{\Delta x \to 0} (3x_0^2 + 3(\Delta x)x_0 + (\Delta x)^2) \qquad (2.2)$$

$$= 3x_0^2 \qquad (2.3)$$

The last two terms are dropped in going from Equation 2.2 to Equation 2.3 because they vanish as Δx approaches zero. We could also just say "$d(x^3)/dx = 3x^2$," leaving out the part which implied that the derivative is actually evaluated at $x = x_0$.

2.1.2 Calculating Derivatives of General Functions

The direct approach to calculating a derivative (explicitly using Equation 2.1) gets quite tedious for more complicated functions, but fortunately it is virtually never necessary. For example, the functions we encountered in the last chapter have quite simple derivatives:

$$\frac{d(x^n)}{dx} = nx^{n-1} \qquad (2.4)$$

$$\frac{d(\sin x)}{dx} = \cos x \qquad (2.5)$$

$$\frac{d(\cos x)}{dx} = -\sin x \qquad (2.6)$$

$$\frac{d(\tan x)}{dx} = \frac{1}{\cos^2 x} \qquad (2.7)$$

$$\frac{d(e^x)}{dx} = e^x \qquad (2.8)$$

$$\frac{d(\ln(x))}{dx} = \frac{1}{x} \qquad (2.9)$$

Incidentally, Equations 2.8 and 2.9 are much simpler than the corresponding equations in other bases:

$$\frac{d(10^x)}{dx} = (2.3023\ldots)(10^x); \quad \frac{d(\log(x))}{dx} = \frac{1}{(2.3023\ldots)x} \qquad (2.10)$$

The extra factor of $\ln 10 = 2.3023\ldots$ in these equations makes base-10 much less convenient. The very simple relationship between an exponential with base e and its derivative is the reason that base is so important, even though $e \approx 2.7183\ldots$ is an irrational number.

Four general relations are widely used to calculate derivatives of more complicated functions. In the next few equations, $f(x)$ and $g(x)$ are two possibly different functions of x, and C is any numerical constant. All of these relations are discussed in the first semester of calculus.

- **Relation 1**: Multiplying a function by any constant multiplies the derivative by the same constant.

$$\frac{d(Cf(x))}{dx} = C\frac{df(x)}{dx} \tag{2.11}$$

Examples:

   a) $d(3x^3)/dx = 3d(x^3)/dx = 9x^2$
   b) $d(6\sin\theta)/d\theta = 6\cos\theta$

- **Relation 2**: The sum of two functions has a derivative that is equal to the sum of the two derivatives.

$$\frac{d(f(x)+g(x))}{dx} = \frac{df(x)}{dx} + \frac{dg(x)}{dx} \tag{2.12}$$

Examples:

   a) $d(x^3 + x^2)/dx = d(x^3)/dx + d(x^2)/dx = 3x^2 + 2x$
   b) $d(x^2 + C)/dx = d(x^2)/dx + d(C)/dx = 2x + 0 = 2x$

The second example above illustrates an important point: adding a constant to any function does not change its derivative.

- **Relation 3**: The product of two functions has a derivative which is related to the derivatives of the individual functions by the expression

$$\frac{d(f(x)g(x))}{dx} = f(x)\frac{dg(x)}{dx} + g(x)\frac{df(x)}{dx} \tag{2.13}$$

Examples:

   a) To find the derivative $df(x)/dx$ of the function $f(x) = x^2\sin x$ let $f(x) = \sin x$ and $g(x) = x^2$ in Equation 2.13. Then we have:

$$\frac{d(x^2\sin x)}{dx} = (\sin x)\left(\frac{d(x^2)}{dx}\right) + (x^2)\left(\frac{d(\sin x)}{dx}\right)$$
$$= 2x\sin x + x^2\cos x$$

   b) $\dfrac{d((\sin x)/x)}{dx} = \dfrac{\cos x}{x} - \dfrac{\sin x}{x^2}$

c) $\dfrac{d(xe^x)}{dx} = xe^x + e^x$

- **Relation 4**: The derivative of complicated functions can be reduced to the derivatives of simpler function by the **chain rule**:

$$\frac{d(f(y))}{dx} = \left\{ \frac{d(f(y))}{dy} \right\} \left\{ \frac{dy}{dx} \right\} \tag{2.14}$$

The chain rule looks deceptively simple but is extremely powerful. For example:

1. To find the derivative $df(x)/dx$ of the function $f(x) = \sin 2x$, let $y = 2x$ and $f(y) = \sin y$. Then we have:

$$\frac{df(y)}{dy} = \cos y = \cos 2x$$

$$\frac{dy}{dx} = 2$$

$$\frac{d(\sin 2x)}{dx} = 2\cos 2x$$

2. To find the derivative of $f(x) = e^{-Cx^2}$ let $y = -Cx^2$ and $f(y) = e^y$ in Equation 2.14. Then we have:

$$\frac{df(y)}{dy} = e^y = e^{-Cx^2}$$

$$\frac{dy}{dx} = -2Cx$$

$$\frac{d(e^{-Cx^2})}{dx} = -2Cxe^{-Cx^2}$$

2.1.3 Second and Higher Derivatives

The derivative dy/dx of a function $y = f(x)$ is also a function, which in turn has its own derivative. This **second derivative** gives the slope of the tangent curves to dy/dx. It is generally written as d^2y/dx^2 or $f''(x)$. It is calculated by applying the definition of a derivative (Equation 2.1) two separate times. Thus, to find the second derivative of the function $y = x^3$, recall that we showed the first derivative is $3x^2$ (Equation 2.3). Equation 2.4 showed that the derivative of x^2 is $2x$. Equation 2.11 then implies that the derivative of $3x^2$ is $6x$. Therefore, we have

$$\left. \frac{d^2y}{dx^2} \right|_{x=x_0} = \left. \frac{d}{dx}\frac{dy}{dx} \right|_{x=x_0} = \left. \frac{d(3x^2)}{dx} \right|_{x=x_0} = 6x_0 \tag{2.15}$$

Third, fourth, and higher derivatives are defined in a similar way.

2.2 APPLICATIONS OF DERIVATIVES

Why do we care about derivatives in a chemistry course? There are several reasons:

2.2.1 Finding Maxima and Minima

At the local peaks and valleys (called *maxima* or *minima*) of any smoothly varying function, $dy/dx = 0$. You will see formal proofs in calculus class, but the basic reason is simple. The tangent at the peak point or at the bottom of the valley must be horizontal. You can see that this is true by looking at Figure 2.1. More quantitatively, suppose $dy/dx > 0$ (sloping up to the right); then moving to a slightly larger value of x must increase y. Suppose $dy/dx < 0$ (sloping up to the left); then moving to a slightly smaller value of x must increase y. In either case, the initial value of x could not have been the maximum.

We will often derive specific expressions in this book (for example, the speed distribution of a monatomic gas at room temperature), and it is much easier to find the maximum values by differentiation than by graphical methods. Incidentally, in the book *Surely You're Joking, Mr. Feynmann*[1] , Richard Feynmann's frat brothers at MIT are musing over the strangely shaped drafting tools called "French curves," and trying to figure out how the shape is chosen. Feynmann tells them they are special curves—however you hold them, the tangent line to the point at the bottom of the curve is horizontal. Try that one on your more gullible friends . . . .

If the second derivative is positive at a point where the first derivative is zero, the point is a minimum; if the second derivative is negative, the point is a maximum. If the second derivative is also zero, $dy/dx = 0$ does not necessarily imply a maximum or a minimum. For instance, the function $y = x^3$ has $dy/dx = 0$ at $x_0 = 0$.

2.2.2 Relations Between Physical Observables

In physics and chemistry, many quantities are directly related by differentiation. We will give only a few examples here:

- The rate of change of position with respect to time is the velocity ($v_x = dx/dt$). The rate of change of velocity with respect to time is acceleration ($a_x = dv_x/dt = d^2x/dt^2$). Thus calculus provides a very natural language for describing motions of particles, and in fact Newton's famous *laws of motion* are best expressed in terms of derivatives (he was one of the inventors of calculus as well). We will discuss Newton's laws in Chapter 3.

- Current, charge, and voltage are extremely important quantities in electrical circuits, including those formed in electrochemical experiments. The rate of change of total charge Q with respect to time is the current ($I = dQ/dt$). The voltage

---

[1]A must-read for any science student—it is the autobiography of Richard Feynmann, Nobel laureate in Physics and one of the most interesting personalities in twentieth century science.

across a capacitor, such as two metal plates separated by air, is proportional ι the charge ($V = Q/C$, where C is the capacitance). The voltage across a resistor is proportional to the current ($V = IR$). The voltage across an inductor, such as a coil of wire, is $V = -L \, dI/dt = -L \, d^2Q/dt^2$, where L is called the *inductance*.

2.2.3 Kinetics of Chemical and Radioactive Processes

The rate of change of concentration of some species A (e.g., $d[A]/dt$) is a measure of the rate of chemical reaction. Most of the equations in chemical kinetics are ***differential equations*** meaning they involve at least one derivative. In addition, nuclear disintegrations (such as $^{14}C \longrightarrow {}^{14}N + e^-$) use the same rate equations as do many unimolecular decompositions (such as $N_2O_4 \longrightarrow 2NO_2$).

In the nuclear case, one electron is emitted for each ^{14}C that disintegrates. The total number of emitted electrons per unit time is proportional to the number of ^{14}C present at that time ($N(t)$, as we defined it in Section 1.4).

$$\text{\# of disintegrations per unit time} = -\frac{dN(t)}{dt} = kN(t) \qquad (2.16)$$

Note the minus sign. Each disintegration decreases the total number of carbons. One solution to this differential equation is apparent from Equation 2.8, which shows that the derivative of an exponential is proportional to itself. You can readily use Equation 2.8 to verify that

$$N(t) = N(0) \exp(-kt) \qquad (2.17)$$

is a solution to Equation 2.16; in fact it is the only solution.

2.2.4 Quantum Mechanics

We will discuss ***quantum mechanics*** extensively in Chapters 5 and 6. It provides the best description we have to date of the behavior of atoms and molecules. The Schrödinger equation, which is the fundamental defining equation of quantum mechanics (it is as central to quantum mechanics as Newton's laws are to the motions of particles), is a differential equation that involves a second derivative. In fact, while Newton's laws can be understood in some simple limits without calculus (for example, if a particle starts at $x = 0$ and moves with constant velocity v_x, $x = v_x t$ at later times), it is very difficult to use quantum mechanics in any quantitative way without using derivatives.

2.2.5 Approximating Complicated Functions

Another common application of derivatives is to generate a simple approximation to some complicated function $f(x)$. We can rearrange Equation 2.1 and substitute $y =$

$$f(x_0 + \Delta x) = f(x_0) + (\Delta x)\frac{df(x)}{dx}\bigg|_{x=x_0} \quad (\text{limit } \Delta x \to 0)$$

If Δx is not infinitesimal, but is "sufficiently small," it must be true that

$$f(x_0 + \Delta x) \approx f(x_0) + \Delta x\frac{df(x)}{dx}\bigg|_{x=x_0} \quad (2.18)$$

Equation 2.18 is actually the first term in what is known as a *Taylor series*, which can be extended to include higher derivatives as well for a better approximation. The more general expression is:

$$f(x_0 + \Delta x) = f(x_0) + \Delta x\frac{df(x)}{dx}\bigg|_{x=x_0} + \frac{(\Delta x)^2}{2}\frac{d^2 f(x)}{dx^2}\bigg|_{x=x_0}$$
$$+ \cdots \frac{(\Delta x)^n}{n!}\frac{d^n f(x)}{dx^n}\bigg|_{x=x_0} + \cdots \quad (2.19)$$

where $n!$ (called "n factorial") $= 1 \cdot 2 \cdot 3 \ldots n$ is the product of all integers up to n if $n > 0$; $0! = 1$. You may have encountered this expression in high school algebra; it also gives the coefficients in front of the different terms in the expansion of $(a + b)^n$. This equation is only useful if the terms eventually get progressively smaller. Fortunately, many simple functions converge quickly to their Taylor expansions. As an example, from Equation 2.8, the derivative of e^x is also e^x, so $d(e^x)/dx|_{x=0} = 1$. Therefore, if $x_0 = 0$ in Equation 2.18,

$$e^x \approx 1 + x \quad (x \ll 1) \quad (2.20)$$

where we have replaced "Δx" with "x" for simplicity because we will not be looking at the limit as this term approaches zero. The full expression in Equation 2.19 is very easy to find, because all of the higher derivatives of e^x are also e^x:

$$e^x = 1 + x + \frac{x^2}{2} + \frac{x^3}{6} + \cdots + \frac{x^n}{n!} + \cdots \quad (2.21)$$

The ratio of the nth term in Equation 2.21 to the immediately preceding term is $x/(n - 1)$. So if $x \ll 1$, each term is much smaller than the one before it, and the series converges rapidly. For example, if $x = 0.01$, $e^{0.01} = 1.010050167$, which differs only slightly from the value of 1.01 predicted by Equation 2.20.

If $x > 1$ the series starts out with growing terms, but no matter how large a number we choose for x, $x/n \ll 1$ for large enough n. Thus eventually the terms start getting progressively smaller. In fact this series converges for all values of x.

Taylor series for the sine and cosine function are also often useful:

$$\sin x = x - \frac{x^3}{3!} + \frac{x^5}{5!} \cdots + (-1)^n \frac{x^{2n+1}}{(2n+1)!} + \cdots \quad (2.22)$$

$$\cos x = 1 - \frac{x^2}{2!} + \frac{x^4}{4!} \cdots + (-1)^n \frac{x^{2n}}{(2n)!} + \cdots \qquad (2.23)$$

Notice that the expansion for the cosine contains only even powers of x, which is expected since $\cos(-x) = \cos x$, and this is only true for even powers; the expansion for the sine contains only odd powers of x.

Sometimes the Taylor series only converges for a specific domain of x values. For example, the Taylor series expansion of $\ln(1 + x)$ (Problem 2-7) only converges for $|x| < 1$. There are even bizarre functions which do not converge at all to their Taylor series expansions. In practice, however, such pathological cases are almost never encountered in physics or chemistry problems, and Taylor series expansions are a very valuable tool.

Another example is useful for simplifying polynomials:

$$1/(1 + x)^n \approx 1 - nx \quad (x \ll 1) \qquad (2.24)$$

2.3   PRINCIPLES OF INTEGRATION

Just as logarithms and exponentials are inverse operations, *integration* is the inverse of differentiation. The integral can be shown to be the area under the curve in the same sense that the derivative is the slope of the tangent to the curve. The most common applications of integrals in chemistry and physics are normalization (for example, adjusting a probability distribution so that the sum of all the probabilities is 1) and calculation of the expectation values of observable quantities.

The curve in Figure 2.2 shows the simple function $y = f(x) = 3x^2$. The area

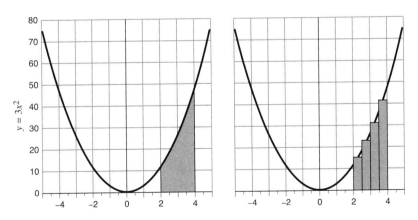

FIGURE 2.2 ▶   Integration gives the area under a curve between two points. **Left**: the shaded area is the integral of $f(x)$ from x_1 to x_2. **Right**: the area can be approximated by adding the areas of a large number of rectangles. This is called *numerical integration*.

under the curve between x_1 and x_2, shaded on the left side of the figure, is written as

$$\int\limits_{x=x_1}^{x_2} f(x)\,dx \tag{2.25}$$

This expression is called "the integral of $f(x)$ from x_1 to x_2." The $\int$ sign signifies an integral. The upper limit is written above the integral sign; the lower limit is below.

One way to approximate the area under the curve would be to replace the complicated shape on the left hand side of Figure 2.2 with the series of rectangles on the right. For example, to get the area between $x = 2$ and $x = 4$, we could break up that range into four boxes, each 0.5 units wide, and each with a height which matches the curve at the middle of the box. The total area of these boxes is 55.875 square units. As we increase the number of boxes, the approximation to the exact shape becomes better, and the total area changes. In this case, with a very large number of boxes, the area approaches 56 square units.

Usually, however, we would prefer to have an explicit functional form for the integral. Since integration is the inverse operation of differentiation, this means that to integrate $f(x)$ we need to find a function whose *derivative* is $f(x)$. This new function is called the **antiderivative**. The difference between the values of this antiderivative function at the two extreme limits of the area gives the value of the integral. For example, we already showed (Equation 2.3) that the derivative of x^3 is $3x^2$, so x^3 is an antiderivative of $3x^2$. Thus we have:

$$\int\limits_{x=2}^{4} 3x^2\,dx = x^3\big|_{x=2}^{x=4} = (4)^3 - (2)^3 = 56 \tag{2.26}$$

The expression "$\big|_{x=2}^{x=4}$" means "evaluated at $x = 4$, minus the value at $x = 2$."

Since the derivative of any constant is zero, the antiderivative can only be determined up to an added constant. For example, the functions $f(x) = x^3$, $f(x) = x^3 + 12$, and $f(x) = x^3 - 3$ all have the same derivative $(df(x)/dx = 3x^2)$. But if you redo the integration in Equation 2.26 using either of these other functions as the antiderivative, you end up with the same answer for the integral. This means that the additive constant can be chosen to be whatever value is convenient, which will be quite important when we consider potential energy in Chapter 3.

One useful general relationship involving integrals is:

$$\int \alpha f(x)\,dx = \alpha \int f(x)\,dx \quad (\alpha \text{ any constant}) \tag{2.27}$$

Thus doubling the height of a curve (setting $\alpha = 2$ on the left side of Equation 2.27) doubles the area under the curve (setting $\alpha = 2$ on the right). We left out the limits of integration, because this result is true no matter what the limits are.

Appendix B lists some commonly used integrals. In general, integration is much harder than differentiation. Fortunately, there are standard reference books with tables of integrals, which fit into two different categories. *Indefinite integrals* give expressions which do not depend on the limits of integration. For example, the first entry in Appendix B is

$$\int x^n \, dx = \frac{x^{n+1}}{n+1}, n \neq -1 \tag{2.28}$$

which simply states that $x^{n+1}/(n+1)$ is an antiderivative of x^n; once again, any constant can be added to the right hand side. To evaluate the integral in Equation 2.26 using this relation, note that Equation 2.28 with $n = 2$ would give the integral of $x^2 (= x^3/3)$. To get the integral of $3x^2$, combine this result with Equation 2.27.

Appendix B also lists some functions integrated for specific limits: these are called *definite integrals*. Perhaps the most important of these functions, which we will use extensively in later chapters, is $f(x) = \exp(-x^2/2\sigma^2)$ (Figure 2.3). This function is called a *Gaussian*. The constant σ adjusts the width of the curve ($f(x)$ is very small if $x \gg \sigma$) and is called the *standard deviation*.

It is possible to prove that

$$\int\limits_{x=-\infty}^{x=\infty} \exp(-x^2/2\sigma^2) \, dx = \sigma\sqrt{2\pi} \tag{2.29}$$

However, no simple function gives the area between arbitrary limits. This integral is so fundamental that its value, numerically integrated by computers to high accuracy, is given its own name—it is called the *error function* and is discussed in Chapter 4.

For tables of integrals see: references [1] and [2].

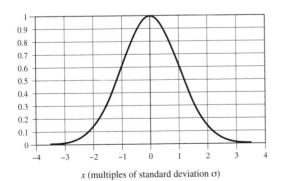

x (multiples of standard deviation σ)

FIGURE 2.3 ▶ The Gaussian function, $f(x) = \exp(-x^2/2\sigma^2)$.

▶ **PROBLEMS** ▶ _____

Calculating Derivatives

2-1.* Find dy/dx if $y = 6 - 2x - x^2$. Use two different approaches—first evaluate this derivative explicitly, as in Equation 2.3; then use the relations listed in Equations 2.4, 2.11 and 2.12.

2-2. Find the value of x which maximizes the function $y = 6 - 2x - x^2$.

2-3.* Use the rules for differentiation in this chapter to find $df(x)/dx$ for the following functions:

(a) $f(x) = \sin^2 x$ (b) $f(x) = \ln(6x)$

2-4. Use the rules for differentiation in this chapter to find $df(x)/dx$ for the following functions:

(a) $f(x) = (\cos x)(\sin x)$ (b) $f(x) = e^{-6x}$

2-5.* Find the first nonzero term in the Taylor series expansion for the function in $(1+x)$. Use this expansion to evaluate ln 1.01, and compare your answer to the exact value.

2-6. Find the error made by the approximation $1/(1 + x)^n \approx 1 - nx$ $(nx \ll 1)$ for $x = 0.01$ and $n = 2$, and for $x = 0.01$ and $n = 50$.

2-7.* Find the full Taylor series expansion for the function $\ln(1 + x)$ (*Hint*: first find a general expression for the nth derivative of the function).

2-8. Use the first term in the Taylor series (Equation 2.18 to prove Equation 2.24.

2-9.* Evaluate the following integrals without using an integral table:

(a) $\int\limits_{x=0}^{x=\pi/2} \sin x \, dx$

Hint: Use Eqns. 2.4 to 2.9 to determine what function has $\sin x$ as its derivative.

(b) $\int\limits_{x=0}^{x=1} e^{2x} \, dx$

Hint: Use Eqns. 2.4 to 2.9 to determine what function has e^{2x} as its derivative.

2-10. Evaluate the following integrals without using an integral table:

(a) $\int\limits_{x=0}^{x=\pi/2} \sin x \, dx$ (b) $\int\limits_{x=0}^{x=1} e^{2x} \, dx$

2-11.* Use Equation B-18 from the table in Appendix B

$$\left(\int\limits_{x=-\infty}^{x=\infty} e^{-x^2/2\sigma^2} \, dx = \sigma\sqrt{2\pi} \right),$$

plus the fact that the function $e^{-x^2/2\sigma^2}$ is symmetric about $x = 0$, to evaluate the integral $\int\limits_{x=0}^{x=\infty} e^{-ax^2} \, dx$ (note the different lower limit).

2-12. Use Appendix B and Equation 2.27 to evaluate the following integrals:

(a) $\int_{x=0}^{x=2\pi} \sin^2 x \, dx$

(b) $\int_{x=-\infty}^{x=\infty} 2\exp(-x^2/8) \, dx$

2-13.* Many chemical species undergo *dimerization reactions*. For example, two molecules of butadiene, C_4H_6, can combine to form the dimer C_8H_{12}. Often such reactions go almost to completion, because the product is more stable than the reactant. Starting with a sample of pure butadiene at time $t = 0$, the concentration of butadiene at a later time t ($[C_4H_6](t)$) is given by the expression:

$$\frac{1}{[C_4H_6](t)} = \frac{1}{[C_4H_6](t = 0)} + kt$$

(a) Find an expression for the rate of change of butadiene concentration $\frac{d[C_4H_6](t)}{dt}$. The correct expression only contains $[C_4H_6](t)$ and k.

(b) How long does it take for the concentration of butadiene to fall to half of its initial value?

2-14. Some chemical reactions obey what is called a "zero-order rate law"—the rate of the reaction is independent of concentration, for a limited time. A typical example might be an enzyme which has a limited number of "active sites," but which binds the reactant so tightly that all the sites are filled if any significant concentration of the reactant is present in solution. Writing the concentration of the reactant as $[A]$, this means that $d[A]/dt = -k$.

(a) Derive an expression for $[A](t)$. Your expression should include the concentration at time $t = 0$ and the rate constant k.

(b) How long does it take for the concentration of the reactant to fall to half its initial value?

2-15. The Environmental Protection Agency has established a guideline for radon concentration in air of 4 picocuries per liter. One curie is defined as 3.7×10^{10} disintegrations per second, so this means one liter of air can have no more than $(4 \times 10^{-12})(3.7 \times 10^{10}) = .148$ radon disintegrations per second. For the isotope of radon most commonly found in basements (^{222}Rn) the half-life $t_{1/2}$ is 3.82 days. Use Equations 1.23 and 2.16 to determine how many radon atoms are in one liter of air which just meets the EPA guidelines, and to determine the concentration of radon in this air (one liter of air at 298K and atmospheric pressure contains about 2.4×10^{22} molecules).

2-16. The unit "curie" used in the last problem is named after Pierre and Marie Curie, who did pioneering experiments with radium in the nineteenth century. One curie (3.7×10^{10} disintegrations per second) is the decay rate of one gram of radium, atomic mass 226 g $\cdot$ mol^{-1}. What is the half-life of radium-226?

Chapter 3

Essential Physical Concepts for Chemistry

Science is divided into two categories: physics and stamp collecting.

Lord Ernest Rutherford (1871–1937)
Nobel Laureate in Chemistry (1908)

Lord Rutherford (who was indeed a physicist, as the quote implies) would have been astonished to see this century's transformation of biology from "stamp collecting" into molecular biology, genomics, biochemistry and biophysics. This transformation occurred only because, time and time again, fundamental advances in theoretical physics drove the development of useful new tools for *chemistry*. Chemists in turn learned how to synthesize and characterize ever more complex molecules, and eventually created a quantitative framework for understanding biology and medicine.

This chapter introduces the core concepts of what is now called *classical physics* (mechanics, electricity, magnetism, and properties of waves). Today we think of classical physics as a special case in a more general framework which would include relativistic effects (for particles with velocities which approach the speed of light) and quantum effects, which are needed for a complete description of atomic behavior. Nonetheless, we will find that this classical perspective (with a few minor corrections) serves as an excellent starting point for understanding many atomic and molecular properties.

3.1 FORCES AND INTERACTIONS

The motions of macroscopic objects (planets or billiard balls) are well described by laws of motion first derived by Sir Isaac Newton in the seventeenth century. Because

Newton's laws long predated relativity and quantum mechanics, we often also call this description *classical mechanics*.

Forces play the central role in the usual formulation of classical mechanics. Forces are vector quantities (they have both direction and magnitude), as discussed in Chapter 1. At a fundamental level, physicists are aware of four distinct forces. The *strong force* and *weak force* are extremely short range, but are responsible for holding nuclei together. The *electromagnetic force* exerted on particle 1 by particle 2 (which we write as $\vec{F}_{1,2}(r)$) is pointed along a unit vector $\hat{r}_{1,2}$ from particle 2 to particle 1 (Figure 3.1). The force depends explicitly on the separation r between the two particles. In SI units, the full expression is

$$\vec{F}_{1,2}(r) = \left\{ \frac{q_1 q_2}{4\pi \varepsilon_0 r^2} \right\} \hat{r}_{1,2} \qquad (3.1)$$

where q_1 and q_2 are the two charges and $\varepsilon_0 = 8.854 \times 10^{-12}$ J^{-1}· C^2· m^{-1} is called the *permittivity of free space*. The extra factor of 4π is explicitly written out (rather than absorbed into the definition of ε_0) to simplify some later equations in electromagnetic theory. Notice that two particles with the same charge give a positive force, which by the definitions in Figure 3.1 is a repulsion. Equation 3.1 is also called *Coulomb's law*, and the force itself is sometimes called the *Coulombic force*. This force is the most important one for describing atomic and molecular interactions.

The final force, the *gravitational force*, is very similar mathematically to the electromagnetic force:

$$\vec{F}_{1,2}(r) = \left\{ \frac{-Gm_1 m_2}{r^2} \right\} \hat{r}_{1,2} \qquad (3.2)$$

where $G = 6.672 \times 10^{-11}$ J · m · kg^{-2} is called the *gravitational constant*. Since masses are always positive, this force is always negative, hence is attractive. The gravitational attraction between a proton and an electron is many orders of magnitude smaller than the Coulombic attraction (Problem 3-1).

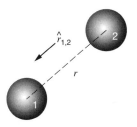

FIGURE 3.1 ▶   The electromagnetic or gravitational force between two particles lies along a vector between the particles.

Forces are important because they affect motion. Motion can be described by the velocity vector $\vec{v}$, or by the momentum vector $\vec{p} = m\vec{v}$. The effects are described quantitatively by **Newton's laws**, which can be paraphrased as follows:

1. In the absence of external forces, the momentum of an object stays constant in direction and magnitude;

2. Force is the derivative of momentum with respect to time. Since $\vec{p} = m\vec{v}$, we can write:

$$\vec{F} = \frac{d\vec{p}}{dt} = \frac{d(m\vec{v})}{dt} = m\frac{d\vec{v}}{dt} = m\vec{a} = m\frac{d^2\vec{r}}{dt^2} \tag{3.3}$$

3. The force exerted on object i by object j (which we will write as $\vec{F}_{i,j}$) is equal in magnitude to the force exerted on object j by object i, and opposite in direction ($\vec{F}_{i,j} = -\vec{F}_{j,i}$).

Equation 3.3 uses the fact that velocity is the derivative of position with respect to time, and acceleration is the derivative of velocity with respect to time. Equation 3.3 is actually a vector equation. If the vectors are expressed in Cartesian coordinates (Section 1.3) it is identical to the three equations

$$
\begin{aligned}
F_x &= \frac{dp_x}{dt} = \frac{d(mv_x)}{dt} = m\frac{dv_x}{dt} = ma_x = m\frac{d^2x}{dt^2} \\
F_y &= \frac{dp_y}{dt} = \frac{d(mv_y)}{dt} = m\frac{dv_y}{dt} = ma_y = m\frac{d^2y}{dt^2} \\
F_z &= \frac{dp_z}{dt} = \frac{d(mv_z)}{dt} = m\frac{dv_z}{dt} = ma_z = m\frac{d^2z}{dt^2}
\end{aligned} \tag{3.4}
$$

Newton's laws are valid in any **inertial frame of reference**. In other words, we can examine the system while we are "at rest," or while we are moving in any direction at a constant velocity. The existence of inertial frames of reference is an assumption, but objects in everyday experience (apples falling from trees, cars colliding on the freeway) satisfy Newton's laws quite well. Strictly speaking, the surface of the Earth does not provide an inertial frame of reference: it rotates about its axis once a day and revolves around the Sun once a year, so our instantaneous velocity is constantly changing its direction. This correction is important for astronomical observations—stars appear to move in circular paths in the night sky, and the forces on them from other stars are far too small for Newton's First Law to be so grossly violated. However, the correction has only minor consequences for molecules or billiard balls.

Some of the consequences of these laws are general and easily derived. In any *closed system*, which may contain a large number of objects but has no forces other than the forces of interaction between these objects, the second and third laws imply that any change to the momentum of object i because of interaction with object j is exactly opposite in direction to the change in momentum of object j. Thus the sum of the momenta of objects i and j is unaffected by the force between them. This can be generalized to show that the total momentum of the system $\vec{p}_{\text{total}}$, which is the sum of all of the individual momentum vectors

$$\vec{p}_{\text{total}} = \vec{p}_1 + \vec{p}_2 + \vec{p}_3 \cdots + \vec{p}_N = \sum_{i=1}^{N} \vec{p}_i \tag{3.5}$$

is unchanged by any of the interactions. We say that the total momentum is *conserved*, or a **constant of the motion**.

3.2 KINETIC AND POTENTIAL ENERGY

Total momentum is not the only quantity that is conserved in a closed system. Another quantity that is conserved is the total energy, but this is a subtler concept than momentum conservation, because energy can be converted between many different forms. Energy can be stored by moving against a force—it takes energy input as work to lift a ball above a table, and then dropping the ball converts this energy input into a velocity. Energy can also leak out of a system in many different ways. For example, friction converts kinetic energy into heat, and the collision of two billiard balls converts a small amount of the kinetic energy into sound.

We will assume in this book that the force depends on only a single coordinate, such as the distance between two particles, and points along that coordinate. Fortunately, this is a very common case. Then we can account for motion against a force by defining a **potential energy function** $U(r)$ such that the derivative of $U(r)$ gives the force:

$$F(r) = -\frac{dU(r)}{dr} \tag{3.6}$$

Note that Equation 3.6 does not use a vector symbol for the force; we already know its direction from the assumption above (to within a sign). Also, the force must be *conservative*, which means in practice that the energy required to move from point A to point B depends only on the two positions, not on other factors such as velocity. Gravity and the force between charges are conservative; friction is not.

Equation 3.6 implies that $U(r)$ is the negative of the antiderivative of $F(r)$, so Equation 3.6 does not uniquely define $U(r)$. A different potential energy function $V(r) = U(r) + C$, where C is *any* numerical constant, would give the same force:

$$\frac{dV(r)}{dr} = \frac{d(U(r)+C)}{dr} = \frac{dU(r)}{dr} + \frac{dC}{dr} = \frac{dU(r)}{dr} \tag{3.7}$$

In other words, adding a constant C to the potential energy function offsets it, but does not change the slope (the derivative). How do we know the "right" value of C to use in order to get the "real" potential energy? We don't. Forces are directly observable (they cause acceleration). *Differences* in potential energy are also observable. Dropping a ball off a table gives a final velocity that depends on the difference in potential energy between the initial and final positions $U(r_{\text{initial}}) - U(r_{\text{final}})$, *but not on either potential energy by itself*, and the unknown numerical constant C does not affect this difference. Thus C has no effect on the time evolution, and is *completely arbitrary*. We usually specify C to make $U(r) = 0$ at some convenient value of r.

Let's explicitly consider a few common cases:

3.2.1 Springs

For an ideal spring

$$U(r) = \frac{k(r - r_0)^2}{2}; \ F(r) = -\frac{dU(r)}{dr} = -k(r - r_0) \tag{3.8}$$

where k is the force constant of the spring and r_0 is the length of the spring when it is neither stretched nor compressed. We will see later that modeling chemical bonds as springs gives a useful description of many molecular properties. The negative sign in the expression for $F(r)$ implies that the force always acts to compress an extended spring, or extend a compressed spring.

3.2.2 Coulomb's Law

The potential energy and force between two charged bodies (in SI units) is given by

$$U(r) = \frac{q_1 q_2}{4\pi\varepsilon_0 r}; \ F(r) = \frac{-dU(r)}{dr} = \frac{q_1 q_2}{4\pi\varepsilon_0 r^2} \tag{3.9}$$

According to this definition of the potential energy, $U(r \rightarrow \infty) = 0$, and the potential energy is negative for oppositely charged bodies at a finite distance ($q_1 q_2 < 0$). Note that a coulomb is a large amount of charge, and it takes a tremendous amount of energy to bring two one-coulomb charges within one meter.

3.2.3 Gravity

The potential energy and the gravitational attractive force between two bodies (masses m_1 and m_2) is given by

$$U(r) = -\frac{Gm_1 m_2}{r}; \ F(r) = -\frac{Gm_1 m_2}{r^2} \tag{3.10}$$

Applying Equations 3.9 and 3.10 for point charges and masses is easy; it is subtler when the sizes of the objects are comparable to their separation. Part of the Earth is immediately underneath your feet, so r is very small from that portion to your body. However, the vast majority of the Earth's mass is thousands of kilometers away. It can be shown that the net gravitational (or Coulomb's law) attraction of a spherical shell on anything outside the shell is the same as if all of the mass (or charge) were concentrated exactly at the center. On the other hand, any object *inside* a spherical shell of charge or mass feels no net force in any direction (Figure 3.2). These effects are very important for multielectron atoms, as will be discussed in Chapter 6; electrons in orbitals that place them close to the nucleus feel very little repulsion from electrons that are farther from the nucleus.

The radius of the Earth at the Equator is 6,378 km, so r in Equation 3.10 changes little as we move a short distance from the surface of the earth. The gravitational force is almost the same on top of Mt. Everest as it is at sea level. This lets us approximate the gravitational attraction to the Earth as:

$$F(r) = -\frac{Gm_1 m_{\text{Earth}}}{r^2} \approx -\frac{Gm_1 m_{\text{Earth}}}{r_{\text{Earth}}^2} = -m_1 g \tag{3.11}$$

where $g = Gm_{\text{Earth}}/r_{\text{Earth}}^2 = 9.8 \text{ m} \cdot \text{s}^{-2}$.

Recall that integration "undoes" differentiation, in the sense described in Chapter 1. This means in turn that we integrate the force to get the difference in potential energy between two positions, say r_1 and r_2:

$$U(r_2) - U(r_1) = -\int_{r=r_1}^{r=r_2} F(r) \, dr \tag{3.12}$$

Equation 3.12 is a form of the **work-energy theorem**. Moving in a direction opposed to a force (such as raising a ball above the earth) requires work, which is stored as potential energy; motion in the same direction as a force (for example, allowing the ball to drop) reduces the potential energy. The right hand side of Equation 3.12 is the work W done on the system. The work-energy theorem states that this work is equal to the

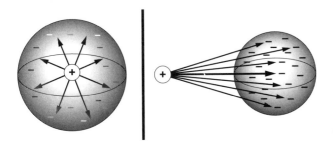

FIGURE 3.2 ▶ A positive charge *inside* a negatively charged spherical shell (left) feels no net force in any direction; the shell has no effect on it. A positive charge *outside* a spherical shell (right) feels a net force that is identical to what it would feel if all of the charge in the shell were concentrated in its center. The same results hold for gravitation attraction.

change in the potential energy of the system (the left-hand side). Again the force must be conservative: rubbing sandpaper against a block of wood requires a work input, but rather than changing the potential energy of the block, the work ends up as dissipated heat.

For example, the work-energy theorem can be used to find the work done by expanding a gas in a piston (Figure 3.3). Suppose $P_{int} > P_{ext}$; recalling the definition of pressure as force per unit area ($P = F/A$), this implies that there will be a net outward force on the piston. Work has to be done on the surrounding gas as the piston is moved outward from $r = r_1$ to $r = r_2$.

$$\begin{array}{l} \text{Work } w \text{ done} \\ \text{on surroundings} \end{array} = U(r_2) - U(r_1)$$

$$= -\int_{r=r_1}^{r=r_2} F(r)\,dr = -\int_{r=r_1}^{r=r_2} (-P_{ext} \cdot A)\,dr \qquad (3.13)$$

We can convert the change in piston position dr into a change in volume dV by noting, from the geometry, that $dV = A\,dr$. Hence

$$w = -\int_{r=r_1}^{r=r_2} (-P_{ext} \cdot A)\,dr = \int P_{ext} \cdot (A\,dr) = \int_{V=V_1}^{V_2} P_{ext}\,dV \qquad (3.14)$$

The internal pressure must exceed the external pressure for any useful work to be done. Remember that result: it is as true for scientists as it is for gases.

Another application of Equation 3.12 is to express gravitational potential energy in a simplified form for objects near the surface of the earth. Since the force in Equation 3.11 is nearly constant, we can approximate the potential energy change for raising

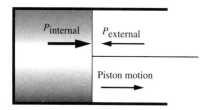

FIGURE 3.3 ▶ A piston with cross-sectional area A will move outward if $P_{internal} > P_{external}$, thereby doing work on the surroundings. The change in position dr becomes a change in volume $dV = A\,dr$. The net work w done on the surroundings is given by Equation 3.14.

an object a distance $h = r - r_{\text{Earth}}$ as:

$$U(r_{\text{Earth}} + h) - U(r_{\text{Earth}}) = -\int_{r=r_{\text{Earth}}}^{r=r_{\text{Earth}}+h} (-mg)\, dr$$

$$= m\, gr\,|_{r_{\text{Earth}}}^{r_{\text{Earth}}+h} = mgh \qquad (3.15)$$

Now the convenient definition is that the potential energy is zero on the surface of the Earth:

$$U(h) = mgh \text{ (near the surface of the earth)} \qquad (3.16)$$

3.3 HARMONIC MOTION

Often Newton's laws predict periodic motion in simple systems. For example, a ball supported from the ceiling by a spring, or a pendulum which is not too far from vertical, will oscillate at a constant and predictable rate. If we connect two balls of comparable mass by a spring and stretch the spring, the entire system will oscillate back and forth, or *vibrate*.

Consider the ball on a spring, which we will assume moves only in the x-direction. If we define $x = 0$ as the position where the spring just counterbalances the force of gravity (so there is no net force or acceleration), we have:

$$F = ma = m\left(\frac{d^2x}{dt^2}\right) = -kx \quad \{\text{from Equations 3.3 and 3.8}\} \qquad (3.17)$$

The second derivative of the position is proportional (with a minus sign) to the position itself. There are only two functions which have this property:

$$d^2\frac{\sin(\omega t)}{dt^2} = -\omega^2\sin(\omega t); \quad \frac{d^2\left(\cos(\omega t)\right)}{dt^2} = -\omega^2\cos(\omega t) \qquad (3.18)$$

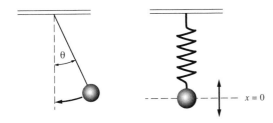

θ

$x = 0$

FIGURE 3.4 ▶ Two systems that undergo simple harmonic motion if displacements are small (left, pendulum; right, mass suspended by a spring).

so the most general solution has the form

$$x = A\cos(\omega t) + B\sin(\omega t) \tag{3.19}$$

where A and B are constants. The velocity and acceleration are found by differentiation:

$$v = -\omega A\sin(\omega t) + \omega B\cos(\omega t)$$
$$a = -\omega^2 A\cos(\omega t) - \omega^2 B\sin(\omega t) = -\omega^2 x \tag{3.20}$$

Combining Equations 3.17 and 3.20 gives

$$\omega = \sqrt{k/m} \quad \text{(harmonic motion, one moving mass)} \tag{3.21}$$

which is the rate of oscillation. Note that the units of ω are radians per second, so the product ωt has units of radians. To convert to a frequency ν (cycles per second) we divide by 2π:

$$\nu = \frac{1}{2\pi}\sqrt{k/m} \tag{3.22}$$

Given some set of initial conditions (the initial position $x(0)$ and the initial velocity $v(0)$), we an determine A and B:

$$x(0) = A;\; v(0) = \omega B \tag{3.23}$$

(Be careful not to confuse the characters for frequency, ν—Greek "nu"— and velocity, v—italic "v".)

The equations above assume that only one end of the spring moves. If both ends can move (as will happen if, for example, the "spring" is really a chemical bond connecting two atoms with similar masses m_1 and m_2) the expression becomes slightly more complicated. If all we are interested in is the relative motion of the two masses, the solutions looks exactly like Equations 3.21 to 3.23 with the mass replaced by $\mu = m_1 m_2/(m_1 + m_2)$.

$$\omega = \sqrt{k/\mu};\; \nu = \frac{1}{2\pi}\sqrt{k/\mu};\; \mu = m_1 m_2/(m_1 + m_2)$$
$$\text{(harmonic motion, two moving masses)} \tag{3.24}$$

The quantity μ has the same dimensions as the mass, and is called the **reduced mass**. Note that if $m_1 \ll m_2$, $\mu \approx m_1$, and Equation 3.24 reduces to Equations 3.21 to 3.23.

3.4   INTRODUCTION TO WAVES

3.4.1 Sound Waves

The collective motions of large numbers of particles often have properties which are only loosely related to the particles themselves, and which are better described using very different concepts. As an example, sound waves are actually waves of gas pressure (Figure 3.5) —the density of gas molecules is alternately slightly higher or lower than

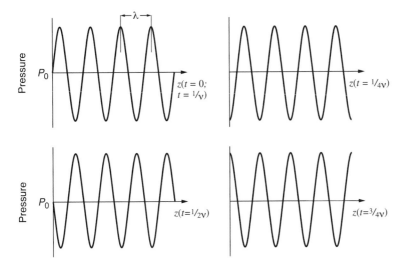

FIGURE 3.5 ▶ Sound waves in air are waves of gas pressure. The positions of the maxima and minima change with time. The waves travel at a characteristic speed $s = \lambda \nu$ determined by properties of the air molecules and independent of wavelength or frequency.

the equilibrium value. The two simplest limiting cases are **plane** *waves*, for which the amplitude varies in only one direction (Figure 3.5) and **spherical** *waves*, which have an origin and spread out in concentric circles from that point. In each case, we can characterize the wave by its **wavelength** (the separation between cycles), which is conventionally called λ. This disturbance travels at a characteristic speed s which is determined by the medium. The energy is dictated by the size of the disturbance (the variation in pressure from maximum to minimum), not by the speed (as it would be for a single particle).

Traveling waves can also be described by their frequency ν, the number of full cycles which pass a given point in one second. Just as in Chapter 1, we can also define an angular frequency $\omega = 2\pi \nu$, the number of radians that pass a given point in one second. The length of one cycle, multiplied by the number of cycles which pass any given point in one second, will give the total distance the wave travels in one second which means

$$\lambda \nu = s \text{ (Wavelength} \times \text{Frequency} = \text{Speed)} \qquad (3.25)$$

The threshold of hearing corresponds to a pressure variation of about 2×10^{-10} atm. Such a wave has a power density of 10^{-20} W $\cdot$m^{-2}. The typical human ear responds comfortably to pressures up to a factor of 10^6 greater than this threshold; at that point, pain is usually felt. The sound waves most young adults can hear have frequencies from roughly 20 cycles per second (20 Hz; very deep bass) to 20,000 cycles per second (20 kHz; very high treble).

The speed of sound in air at room temperature is about 330 m · s^{-1} independent of wavelength. Thus, from Equation 3.25, most young adults can hear sound with wavelengths in air between 16.5 m and 16.5 mm. Small animals can generally hear higher frequencies because the active structures in their ears are smaller; large animals can often hear lower frequencies. The speed of sound is generally much higher in liquids or solids than in gases: sound travels at 1500 m · s^{-1} through water, and about 4000 m · s^{-1} through hardwood.

Musical instruments create tones by restricting the possible wavelengths. For example, a violin string is held rigid at two points, separated by a distance L. The waves generated by drawing the bow across the string must have zeroes at the two rigid points. Since a sine wave goes through zero every half cycle, the only waves consistent with this constraint must satisfy the condition:

$$L = \frac{\lambda}{2}, \lambda, \frac{3\lambda}{2}, \cdots, \frac{n\lambda}{2} \cdots \tag{3.26}$$

or equivalently, the frequency ν is restricted:

$$\nu = \frac{s}{2L}, \frac{s}{L}, \frac{3s}{2L}, \cdots, \frac{ns}{2L} \cdots \tag{3.27}$$

The lowest frequency is called the *fundamental*; all of the other frequencies are multiples of the fundamental and are called *harmonics*. Doubling the frequency corresponds to raising a note by one octave. When a piano and a flute play middle-A, they both produce a distribution of sound waves with a fundamental frequency of 440 Hertz, but they sound different because the amplitudes of the different harmonics depend on the instrument.

3.4.2 Electromagnetic Waves

Electricity and magnetism seem superficially to be as different as a lightning bolt and a compass. However, in the nineteenth century a series of elegant experiments showed electricity and magnetism to be closely related phenomena. Charges create electric fields, even when they are not moving. Moving charges create currents, which in turn generate magnetic fields. These experiments ultimately led to a set of differential equations called *Maxwell's equations*, which provide a completely unified description of the electric and magnetic fields generated by any given charge distribution.

Electric fields and magnetic fields (denoted $\vec{\mathcal{E}}$ and $\vec{B}$ respectively in this book) are vector quantities. A charge q in an electric field feels a force $\vec{F} = q\vec{\mathcal{E}}$. A charge moving at velocity $\vec{v}$ in a magnetic field $\vec{B}$ feels a force of magnitude $\left|\vec{F}\right| = |\vec{v}|\left|\vec{B}\right|\sin\theta$, where θ is the angle between the two vectors, in a direction perpendicular to the plane containing both $\vec{v}$ and $\vec{B}$. The SI unit for electric fields is volts per meter (V/m). The SI unit for magnetic fields is the Tesla (T).

Constant (or **static**) electric and magnetic fields are present nearly everywhere. For example, the two leads of a standard 9V battery are separated by approximately 6 mm at their closest point. To a good approximation, the electric field between the leads points toward the negative lead, and has magnitude $(9V)/(.003 \text{ m}) = 1500$ V/m. This field has no obvious effect on uncharged bodies that do not conduct electricity. The small amount of ions naturally present in air are accelerated by this voltage, but they collide with other molecules and randomize their velocity long before they can build up much excess energy. At substantially higher fields (about 2×10^6 V/m) the excess energy buildup from this acceleration becomes serious and air breaks down, causing a **spark**. In Chapter 8 we will discuss the fields seen by electrons in atoms, which are much higher still (about 10^{11} V/m).

The Earth's magnetic field is approximately 50μT. The position of the magnetic North Pole (currently in extreme northern Canada) wanders somewhat: European navigators in the 15th century found that compasses pointed somewhat east of true north, but in more recent times the deviation has been west. Over geologic time, the Earth's field occasionally becomes small and then changes direction. A very large laboratory magnet would generate a peak field of 15–20T by circulating current through tens of km of coiled superconducting wire cooled to liquid helium temperature (4K). Large magnets are available in essentially every chemistry building, because of their use for nuclear magnetic resonance (NMR), a spectroscopic technique we will discuss in Chapter 5. They are also used in hospitals for magnetic resonance imaging (MRI).

Sound waves cannot travel through a vacuum, where there is no matter to support a density variation. However, electric and magnetic fields can travel though a vacuum. The solutions to Maxwell's equations for such fields show that an oscillating single-frequency electric field will only propagate in free space if there is also an oscillating magnetic field perpendicular to it. The ratio between the electric field and magnetic field amplitudes is fixed, so we refer to this as an **electromagnetic wave**. If the wave is traveling in the z-direction and we choose the x-direction as the direction of the electric field, the general form for a plane electromagnetic wave turns out to be:

$$\vec{\mathcal{E}}(x, y, z, t) = \mathcal{E}_{max}\hat{x} \cos\left(\frac{2\pi z}{\lambda} - \omega t\right) = \mathcal{E}_{max}\hat{x} \cos\left(\frac{2\pi z}{\lambda} - 2\pi v t\right)$$

$$\vec{B}(x, y, z, t) = B_{max}\hat{y} \cos\left(\frac{2\pi z}{\lambda} - \omega t\right) = B_{max}\hat{y} \cos\left(\frac{2\pi z}{\lambda} - 2\pi v t\right) \quad (3.28)$$

$$\frac{\mathcal{E}_{max}}{B_{max}} = c; \quad \lambda v = c$$

where $\hat{x}$ and $\hat{y}$ are unit vectors in the x- and y-directions, respectively. What we call "light" is an electromagnetic wave which can be detected by our eyes, which are sensitive to only a very small distribution of wavelengths—running roughly from $\lambda = 700$ nm (red) to $\lambda = 400$ nm (violet). In a vacuum, the maxima and minima move at the speed of light c. Waves propagating through the atmosphere move a little slower than c; radiation with $\lambda = 590$ nm (yellow light) moves at a speed $s = c/1.0002765$. In general $s = c/n$, where n (called the ***index of refraction***) depends on both the wavelength and the material. At an interface between two materials, the difference in the indices of refraction dictates the amount light changes direction (see Problem 1-15).

Electromagnetic waves transmit energy as well. The average intensity I (in watts per square meter) is proportional to the average squared electric field:

$$I = \frac{\overline{\mathcal{E}^2}}{754} \tag{3.29}$$

The average intensity of sunlight hitting the surface in the continental U.S. at noon is about 200 W/m^2. A typical laser pointer produces 5 mW in a 1 mm^2 spot, so the intensity is about 5000 W/m^2. At the opposite extreme, commercially available "ultrafast laser systems" can readily produce 100 mJ pulses with 100 fs duration (10^{12} W peak power). This peak power is roughly equal to the worldwide electrical generation capacity. Such pulses can be focused to about 10^{-10} m^2 (10^{22} W/m^2); still higher peak powers are created for specialized applications, such as laser fusion.

We can broadly categorize the properties of electromagnetic waves by wavelength (Table 3.1). For example, an AM radio station transmits waves with center wavelength between $\lambda = 545$ m ($\nu = 550$ kHz) and $\lambda = 187$ m ($\nu = 1600$ kHz). The sound is encoded on the electromagnetic wave by making the amplitude of the wave change with time (hence AM for amplitude modulation). As we discuss in Chapter 8, modern chemical laboratories include different instruments that probe molecular responses to radiation all the way from the radio frequency to the X-ray regions.

TABLE 3.1 ▶ **Approximate Regions of the Electromagnetic Spectrum**

Region	λ (meters)	Typical Sources	Molecular effects
Low frequency	$> 10^3$	power lines ($\nu = 60$ Hz, $\lambda = 5000$ km	
Radio frequency	10^3–1	AM/FM radio; television	
Microwaves	1-10^{-3}	radar; microwave ovens	excites rotations
Infrared	10^{-3}–10^{-6}	heated objects	excites vibrations
Visible	4–7×10^{-7}	sun	excites electrons in some molecules
Ultraviolet	2–4×10^{-7}	sun	breaks bonds
X-rays	10^{-7}–10^{-11}	special sources	breaks bonds if absorbed
γ-rays	$< 10^{-11}$	nuclear disintegrations	breaks bonds if absorbed

3.4.3 Properties of Waves

Waves have properties that are often quite different from what we would associate with billiard balls or other particles. These differences will become very important when we discuss quantum mechanics.

	Particles (baseball) versus	Waves (water wave)
Position	localized	delocalized
Interference	no	yes
Energy (free space)	$\propto (\text{velocity})^2$	$\propto (\text{amplitude})^2$
Speed	variable from 0 to 3×10^8 m/sec, continuously	fixed by medium (e.g., sound in air, 330 m/sec)
Mass	yes	no
Momentum	yes, $\vec{p} = mv^2$	yes; water waves smash buildings

Probably the most dramatic difference in the behavior of particles and waves is the possibility of ***interference*** for waves when multiple sources are present. Figure 3.6 illustrates the differences. The top of Figure 3.6 illustrates purely particle-like behavior (for instance, firing two shotguns simultaneously). The particles from each shotgun

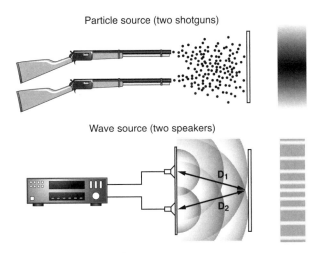

Particle source (two shotguns)

Wave source (two speakers)

D_1

D_2

FIGURE 3.6 ▶ Comparison of the total distribution produced by two particle sources or two wave sources. Notice that waves exhibit interference, so the amplitude at some positions is far lower than the amplitude which would be produced by either source alone.

have a range of velocities and directions, so each shotgun gives a spread of positions on a faraway target. Suppose these spreads overlap, but the probability of two pellets colliding in midair is small; then the total distribution of pellets on the target is obviously just the sum of the spreads produced by each shotgun individually.

The bottom of Figure 3.6 illustrates purely wavelike behavior (for instance, playing the same tone through two speakers). When two waves overlap, their maxima can reinforce (causing *constructive interference* and increasing the amplitude of the disturbance) or can cancel (causing *destructive interference* and decreasing its amplitude). We usually see constructive interference in some directions and destructive interference in others. If the paths to the two sources differ by an integral number of wavelengths $(D_1 - D_2 = N\lambda)$ the maxima and minima of the two waves will coincide and reinforce. If they differ by a half-integral number of wavelengths, the maxima from one source will coincide with the minima from the other, and the waves will cancel.

The separation between maxima (often called *fringes*) depends explicitly on the wavelength, so interference can be used to measure wavelength. For example, a *diffraction grating* consists of a series of regularly spaced dark lines. Light passes through the spaces between the lines, so we see constructive and destructive interference as if there were many wave sources at slightly different positions. If light comes in perpendicular (*normal*) to the grating, as in Figure 3.7, some of the intensity will come out in different directions θ. For the waves to reinforce in any particular direction, the path length differences between the rays that hit different spaces (shown here as a dashed line) must be an integral number of wave lengths. Geometrical arguments

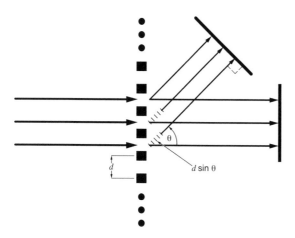

FIGURE 3.7 ▶ Magnified view of light hitting a diffraction grating at normal incidence. Much of the light just goes through ($\theta = 0$). For any other direction θ, the light from the different spaces in the grating travels along paths with different lengths. The path length difference (shown here as a dashed line) is $d \sin \theta$. If this is an integral number of wavelengths ($n\lambda = d \sin \theta$) waves along these paths can constructively interfere, and this generates an increased intensity.

show that this length is $d \sin \theta$ so we have

$$n\lambda = d \sin \theta \quad \text{(diffraction grating, normal incidence)} \qquad (3.30)$$

Equation 3.30 is called the ***diffraction equation***. The solution with $n = 0$ ($\theta = 0$) is simple transmission of the light. The solutions for $n \neq 0$ give intensity in other directions, and the positions of these additional spots can be used to determine λ if d is known. Thus, optical scientists can use a manufactured diffraction grating with known line separations to measure the wavelength of an unknown light source.

X-rays have wavelengths that are comparable to the spacing between atoms. Rows of atoms cause diffraction, so if the wavelength of the X-rays is known, the spacing between atoms can be determined. Figure 3.8 shows that the condition for constructive interference, hence strong scattered X-rays, is

$$N\lambda = 2d \sin \theta \quad \text{(diffraction off a crystal lattice)} \qquad (3.31)$$

This technique, called ***X-ray crystallography***, can be extended to measure the detailed structures of even complicated materials such as proteins with literally thousands of atoms. However, it is much harder to infer molecular structure if the material cannot be grown as a single crystal, and crystal growth can be exceedingly difficult.

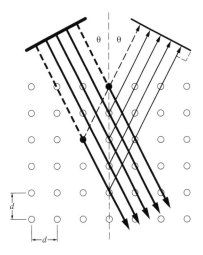

FIGURE 3.8 ▶ X-rays diffracted off a crystal (viewed here as a regular array of atoms) will constructively interfere if multiple scattering paths reinforce. For example, in this picture, the two black atoms both produce scattered X-rays which would superimpose. The difference in the total path length (along the dashed paths) would be $2d \sin \theta$. So the constraint for constructive interference is $N\lambda = 2s \sin \theta$.

3.5 INTRODUCTION TO ATOMIC AND MOLECULAR INTERACTIONS

3.5.1 Chemical Bonds

When atoms combine to form molecules, they create chemical bonds which can be described by a potential energy function. An accurate description of the bonding requires quantum mechanics, as we discuss in Chapters 6 and 8, but many of the features can be understood with the "classical" picture we have been developing in this chapter.

Bonds are broadly grouped into two types: ionic (for example, in the molecule KCl, which can be written to a good approximation as K^+Cl^-) and covalent (for example, in homonuclear diatomic molecules such as O_2 or I_2). We will consider the covalent case first. As two neutral atoms grow closer, the potential energy decreases until a minimum is reached; at still shorter distances, the potential energy eventually becomes positive. Figure 3.9 shows the actual, experimentally measured potential energy for two iodine atoms (solid line); the bottom of the potential well, which corresponds approximately to the bond length, is 2.5×10^{-19} Joules below the energy as $r \to \infty$. It is more common to multiply this value by Avogadro's number, thus giving a well depth of 150 kJ/mol in this case.

The exact shape of the potential energy curve is different for each possible pair of atoms, and can only be calculated by a detailed quantum mechanical treatment. One convenient approximate potential for covalent bonds is the ***Lennard-Jones 6–12 potential***, shown as the dashed line in Figure 3.9 and in more detail in Figure 3.10. The equation for this potential is:

$$U(r) = 4\varepsilon \left(\left(\frac{\sigma}{r}\right)^{12} - \left(\frac{\sigma}{r}\right)^{6} \right) \tag{3.32}$$

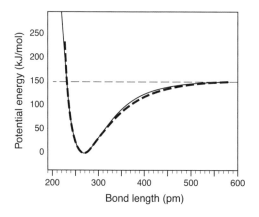

FIGURE 3.9 ▶ Comparison of the actual interatomic potential for two iodine atoms (solid line) with the approximate Lennard-Jones potential, Equation 3.32. The Lennard-Jones potential reproduces the important features.

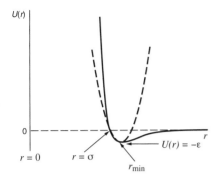

FIGURE 3.10 ▶ Comparison of the Lennard-Jones potential with a still simpler approximation: a parabola with the same minimum and second derivative at the minimum. Near the bottom of the well, this potential (the same as a spring) provides a good approximation.

As $r \to \infty$, $U(r) \to 0$; as $r \to 0$, $U(r) \to \infty$. At $r = \sigma$, $U(r) = 0$. The minimum energy position will be where $dU(r)/dr = 0$:

$$\left. \frac{dU(r)}{dr} \right|_{r=r_{min}} = 4\mathcal{E}\left(\frac{-12\sigma^{12}}{r_{min}^{13}} + \frac{6\sigma^6}{r_{min}^7} \right) = 0$$

$$r_{min} = (2)^{1/6}\sigma; \quad U(r_{min}) = -\mathcal{E} \tag{3.33}$$

The position of the potential minimum r_{min} and the well depth $\mathcal{E}$ are listed for a variety of diatomic molecules in Table 3.2, and these parameters can be used to construct an approximate potential using Equation 3.32. This equation also can be used to evaluate the force along the internuclear axis, since $F = -dU/dr$. When $r > r_{min}$, the force is attractive because $dU/dr > 0$; when $r < r_{min}$ the force is repulsive.

The Taylor series (discussed in Chapter 2) lets us simplify any potential further in the region close to the minimum r_{min}. Saving the terms through the second derivative gives

$$
\begin{aligned}
U(r) &\approx U(r_{min}) + (r - r_{min})\left. \left(\frac{dU(r)}{dr} \right) \right|_{r=r_{min}} \\
&\quad + \left(\frac{(r-r_{min})^2}{2} \right)\left. \left(\frac{d^2U(r)}{dr^2} \right) \right|_{r=r_{min}} \\
&= U(r_{min}) + \left(\frac{(r-r_{min})^2}{2} \right)\left. \left(\frac{d^2U(r)}{dr^2} \right) \right|_{r=r_{min}}
\end{aligned}
\tag{3.34}
$$

since the first derivative vanishes at the minimum. Equation 3.34 has the same form as the potential energy of a spring (Equation 3.8) with the zero of energy chosen as the bottom of the potential well. Thus, for energies near the bottom of the well, modeling the internuclear interaction as if there were a "spring" connecting the atoms is often a good approximation (Figure 3.10). Comparing Equations 3.8 and 3.34 gives an effective "force constant" $k = d^2U(r)/dr^2\big|_{r=r_{min}}$ for the chemical bond. Experimental values are listed in Table 3.2.

TABLE 3.2 ▶ Chemical Bond Characteristics for Selected Diatomic Molecules

Molecule	Experimental Dissociation Energy (kJ · mol⁻¹)	Potential Well Depth $\mathcal{E}$ (kJ · mol⁻¹)	Separation r_{min} at Potential Minimum (pm)	Force constant (N · m⁻¹)
H_2	432	458	74	570
HF	565	590	92	970
HCl	428	445	128	520
HBr	364	380	141	410
HI	295	308	160	310
I_2	149	150	267	170
Cl_2	239	242	199	330
ICl	207	210	232	240
O_2	494	512	121	1180
N_2	941	955	109	2300
CO	1072	1085	113	1902
Ionic bonds:				
NaCl	552*	554*	236	109
KCl	490*	491*	267	109

*energy below separated ions; all other energies are relative to neutral atoms

The potential energy function looks somewhat different for an ionic bond, such as in the molecule $^{39}K\ ^{37}Cl$, which can be approximately written as the ion pair $^{39}K^+$——$^{37}Cl^-$. Those two ions have the same number of electrons and neutrons, and differ only in the number of protons. The chloride ion is larger (181 versus 133 pm) because the natural repulsion of the electrons is counteracted by two fewer protons. At long distances the charge distributions are symmetrical, so as shown in Figure 3.11 we can replace the electrons and protons with a single positive charge for K^+, and a single

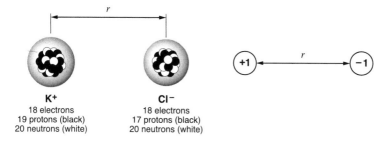

K⁺
18 electrons
19 protons (black)
20 neutrons (white)

Cl⁻
18 electrons
17 protons (black)
20 neutrons (white)

FIGURE 3.11 ▶ For large separations, the complicated interaction between the 37 charged particles in K^+ and the 35 charged particles in Cl^- reduces to simple Coulombic attraction between the two net charges. Effects from the spherical electronic clouds and the small nucleus tend to cancel (see Figure 3.2).

negative charge for Cl^-. Thus at long distances the potential energy is proportional to $1/r$ (Coulomb's law), which is a much slower falloff than the $1/r^6$ dependence in the Lennard-Jones potential.

In addition, even though as chemists we think of K^+ as more stable than K and Cl^- as more stable than Cl, the ion pair K^+–Cl^- is less stable than the pair of neutral atoms at infinite separation. The energy difference is quantified by the *ionization potential* (IP) and the *electron affinity* (EA), which are defined for an arbitrary atom A as:

$$A \rightarrow A^+ + e^- \quad \text{energy difference } \Delta E = I P$$
$$A^- \rightarrow A + e^- \quad \text{energy difference } \Delta E = E A \qquad (3.35)$$

Some typical values are listed in Table 3.3.

TABLE 3.3 ▶ Electron Affinities and Ionization Potentials for Selected Neutral Atoms

Atom	Ionization Potential (kJ · mol^{-1})	Electron Affinity (kJ · mol^{-1})
H	1312	73
Li	520	60
C	1086	122
F	1681	328
Na	496	53
Cl	1255	349
K	419	48

Notice that all of these numbers are positive. Thus, for example, it takes 419 kJ · mol^{-1} to remove an electron from a neutral potassium atom; it takes 349 kJ · mol^{-1} to remove an electron from Cl^-. This implies that at infinite separation, the reaction $K + Cl \rightarrow K^+ + Cl^-$; requires an energy input:

$$\Delta E = I P(\text{K}) - E A(\text{Cl}) = +70 \text{ kJ } \cdot \text{ mol}^{-1}$$

The molecule gains back this energy (and more) due to the Coulombic attraction as the atoms move from infinite separation to the experimentally observed bond distance of 267 pm. Coulombic attraction would tend to draw the two ions as close as possible, but we will see later (in Chapters 5 and 6) that quantum mechanics predicts the energy will eventually start to rise if the atoms get too close. Combining all of these concepts gives a commonly used approximate potential for ionic bonds of the form

$$U(r) = Ae^{-Br/r_{\min}} - \frac{e^2}{4\pi \varepsilon_0 r} + I P - E A \qquad (3.36)$$

The last two terms reflect the energy difference between the ion pair and neutral atoms.

3.5.2 Diatomic Molecules—Degrees of Freedom

Some molecular properties can be understood by picturing molecules as a collection of masses connected by massless springs (Figure 3.12). For a diatomic molecule, the motions of the two masses (m_1 and m_2) can be completely described by six components: the x-, y- and z-components of the velocities of each mass. However, these motions are coupled. For example, if mass 1 in Figure 3.12 is moving towards mass 2 and mass 2 is motionless, the spring will shrink; this in turn will supply a force to mass 2, and energy will oscillate back and forth between the kinetic energy of the two masses and potential energy in the spring.

These six distinct velocities or any combinations of them are called ***degrees of freedom***. It is possible to choose combinations of velocities on the different atoms in a way which simplifies the subsequent motion. For example, if the spring is initially at its rest length and if mass 1 and mass 2 are both moving at the same velocity, the separation between the masses does not change and the masses continue to move at their initial velocities for an indefinite time. In general we can group the possible motions of the two masses into three different categories: translation, rotation, and vibration.

1. If both of the masses are moving in the same direction at the same speed, the system is ***translating***. This motion does not change the separation between the two masses so it has no effect on the spring. In effect, the translation describes the characteristics of the two masses taken as a single, structureless "object." In deriving the kinetic theory of gases (Chapter 7) we will use this translational energy, and the momentum it implies, to calculate pressure.

    Translation is completely described by three degrees of freedom—motion of the entire system in the x-, y-, or z- direction (the three motions on the top of Fig-

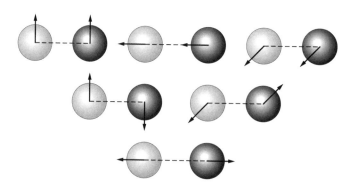

FIGURE 3.12 ▶ Two masses can be described by six distinct motions, or degrees of freedom. The motions can be grouped logically into translations of the two masses together (top), rotations (middle) and a single vibration (bottom). In the absence of collisions, the six motions drawn here are completely separate from each other; energy does not flow between rotations, vibrations, and translation.

ure 3.12). Since total momentum is conserved in the absence of external forces (such as collisions), these three velocity components will remain constant.

2. Suppose we choose a frame of reference where the total momentum is zero, and suppose further that the separation r between the masses is fixed (for example, by a rigid rod). The masses could still be *rotating* at some frequency ω about their center of mass. If the spring is initially pointed along the x-direction, then the axis of rotation can be in the y-direction, the z-direction, or any combination of the two (the middle row of Figure 3.12). Thus there are two degrees of freedom associated with rotation in a two-mass system.

Rotational energy contributes to the internal energy of a diatomic molecule, and classically any rotational speed is possible. We will return to rotational properties in Chapter 8, when we discuss quantum mechanics, which imposes restrictions on the rotational energy; we will find that transitions between allowed rotational states let us measure bond lengths or cook food in microwave ovens.

3. Finally, suppose the masses are not translating (the total momentum is zero) or rotating. This means the components of the velocities perpendicular to the spring are zero. However, it is still possible for the length of the spring to change because of forces generated by the internuclear potential discussed in the last section. Then the system is *vibrating* (the bottom row of Figure 3.12). For small vibrational energy, the rate of the vibration will be given by Equation 3.24

$$\omega = \sqrt{\frac{k}{\mu}}$$

where k is the force constant of the spring connecting the two masses and $\mu = m_1 m_2 / (m_1 + m_2)$ is the reduced mass.

Consider, for example, carbon monoxide. The mass of one mole of carbon-12 atoms is exactly 12 g; dividing by Avogadro's number (and converting to kg) gives the mass of a single carbon-12 atom as 1.9926×10^{-26} kg. The mass of one mole of oxygen-16 atoms is 15.9949 g, so the mass of one atom is 2.6560×10^{-26} kg (masses in amu for many different isotopes are listed in Appendix A). The reduced mass $\mu = m_C m_O / (m_C + m_O)$ is then 1.1385×10^{-26} kg.

We will show in Chapter 8 that the vibrational motion of atoms in molecules is responsible for the **greenhouse effect** which tends to increase the Earth's temperature. We will also show that the modern (quantum mechanical) picture does not permit the molecule to merely sit with its atoms separated by the minimum potential energy. Even at absolute zero, the molecule still has total energy $E = \hbar\omega/2$, where $\hbar = 1.054 \times 10^{-34}$ J · s (see Appendix A). So the actual dissociation energy is always less than the depth of the potential well, as shown in Table 3.2.

The three translations, two rotations, and one vibration provide a total of six independent motions. In the absence of external forces (such as collisions), energy is never exchanged between translation, rotation and vibration, and translational motion in the x-direction is never changed into translational motion in the y- or z-directions. So we can treat each motion independently. Thus this decomposition into translation, rotation and vibration simplifies matters considerably.

3.5.3 Polyatomic Molecules

This separation of the different motions into translation, rotation and vibration can also be generalized to polyatomic molecules. N atoms have $3N$ degrees of freedom (motion of each atom in the x-, y-, or z- directions). These motions can be decomposed into translation, rotation and vibration as well. If we treat the collection of atoms as a rigid, nonrotating "object," it can move as a whole in the x-, y- or z-directions: thus there are still three translational degrees of freedom. If the molecule is linear, there are only two rotational degrees of freedom, as in the diatomic case (rotation about the axis of the spring does not change anything at all, but rotation about any axis perpendicular to the spring does have an effect). If the molecule is nonlinear, rotation about all three directions in space is different. Thus a nonlinear molecule has six rotations and translations; a linear molecule has five rotations and translations. This implies:

  Linear N-atom molecule: 3 translations, 2 rotations, $3N - 5$ vibrations
  Nonlinear N-atom molecule: 3 translations, 3 rotations, $3N - 6$ vibrations

Table 3.4 gives typical bond energies and force constants for bond stretches (averaged over a large number of different molecules), but many of the vibrational modes in large molecules have much smaller restoring forces. For a general polyatomic molecule, there are usually more vibrations than bonds, and finding stable vibrational excitations (called **normal modes**) can be a quite complicated task. Figure 3.13 shows the three

TABLE 3.4 ▶ Typical Chemical Bond Characteristics (averaged over many molecules)

Type of Bond	Dissociation Energy $(kJ \cdot mol^{-1})$	Separation at potential minimum (pm)	Force const. $(N \cdot m^{-1})$.
H–C	413	110	500
H–N	391	101	597
H–O	467	96	845
C–C	347	154	550
C=C	614	134	843
C≡C	839	120	1500
C–O	358	143	540
C=O	799	123	1600
C–N	305	143	512
C≡N	891	116	1850

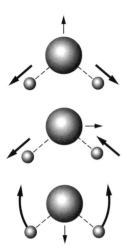

FIGURE 3.13 ► The three normal vibrational modes of water. For the top mode (the *symmetric stretch*) both O–H bonds are extended or compressed at the same time. For the middle mode (the *antisymmetric stretch*) one O–H bond is extended when the other is compressed. The bottom mode is called the bend. In every case the hydrogen atoms move more than the oxygen, because the center of mass has to stay in the same position (otherwise the molecule would be translating). For a classical molecule (built out of balls and perfect springs) these three modes are independent. Thus, for example, energy in the symmetric stretch will never leak into the antisymmetric stretch or bend modes.

normal vibrational modes of water. Note that, for example, just stretching one of the O–H bonds in water does not produce a normal mode; that would force two of the atoms to move, which would in turn induce a force on the other hydrogen atom.

The two stretching modes have nearly the same vibrational frequency (110 versus 113 THz) but the bending mode frequency is much lower (48 THz), reflecting the lower restoring force for the bending motion. In larger molecules, some of the internal motions have very low restoring forces. For example, in the ethane molecule (H_3C–CH_3) rotation of the two –CH_3 groups about the central C–C bond is essentially unhindered.

3.5.4 Intermolecular Interactions

The network of chemical bonds within molecules only partially determines the chemical and physical properties. Intermolecular interactions cause gases to deviate from the ideal gas law (as we will discuss in Chapter 7) or condense into liquids. Interactions between nonbonded atoms in the same molecule cause proteins to fold into specific configurations, which catalyze chemical reactions critical to life.

Interactions between neutral, nonpolar atoms or molecules are relatively weak, and can be accurately modeled by the Lennard-Jones potential discussed in the last section. Table 3.5 lists some specific examples. Notice that the well depth is less than 1% of typical bond energies in Table 3.2. In fact none of these atoms and molecules is condensed into a liquid at STP (standard temperature and pressure; $P = 1$ atm, $T = 273$K).

TABLE 3.5 ▶ **Lennard-Jones Parameters for Intermolecular Interactions**

Molecule or Atom	ε	σ [pm]
He	0.14×10^{-21} J (0.085 kJ $\cdot$ mol^{-1})	256
Ne	0.49×10^{-21} J (0.30 kJ $\cdot$ mol^{-1})	275
Ar	1.66×10^{-21} J (1.00 kJ $\cdot$ mol^{-1})	341
Kr	2.26×10^{-21} J (1.36 kJ $\cdot$ mol^{-1})	383
Xe	3.16×10^{-21} J (1.90 kJ $\cdot$ mol^{-1})	406
CH_4	2.05×10^{-21} J (1.24 kJ $\cdot$ mol^{-1})	378
$C(CH_3)_4$ (neopentane)	3.20×10^{-21} J (1.93 kJ $\cdot$ mol^{-1})	744
CO_2	2.61×10^{-21} J (1.57 kJ $\cdot$ mol^{-1})	449

Intermolecular interactions are much stronger in molecules which have a partial charge separation (***polar*** molecules). For example, the electrons in a hydrogen–oxygen bond tend to be pulled more towards the oxygen atom (we say that oxygen is more ***electronegative*** than hydrogen), resulting in an electric ***dipole moment*** (Figure 3.14, left).

Electric dipoles interact much more strongly than do neutral molecules (at long range, the interaction is proportional to $1/r^3$ instead of $1/r^6$). Probably the most important example of a dipole-dipole interaction is the ***hydrogen bond***, which generally has the form X–H$\cdots$Y, where X and Y are electronegative elements such as N, O, or F, and the X–Y dipole interacts with a lone pair of electrons on the Y atom. The right side of Figure 3.14 shows the structure of the gas-phase water dimer. Hydrogen bonds have a typical strength of 25–50 kJ $\cdot$ mol^{-1}. They dramatically stabilize liquid water (each water molecule has two hydrogens and two lone pairs of electrons, so it can participate in two hydrogen bonds). They also hold together DNA strands in a "double helix," and account for much of the stability in protein folding.

Polar molecules can also serve to stabilize ionic charges in solution, as illustrated schematically in Figure 3.15. Anions are surrounded by the hydrogen end of the water molecules; cations are surrounded by the oxygen end. This serves to partially shield the charges in solution, thus reducing the energy associated with charge separation. As a result, salts dissolve more readily in polar solvents than in nonpolar solvents.

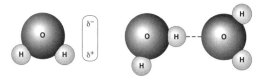

FIGURE 3.14 ▶ **Left**: The electrons in water tend to be pulled more towards the oxygen atom, resulting in an electric dipole moment. **Right**: Structure of the water dimer in the gas phase, showing a typical hydrogen bond. The complex rotates freely about the dashed line; the atoms do not always lie in a single plane.

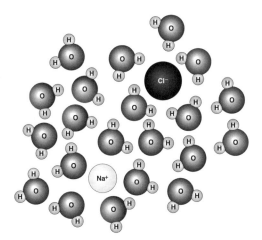

FIGURE 3.15 ▶ Schematic illustration of the structure of a salt solution. The cations are attracted to the negative charge density on the oxygen, which increases stability; the anions are attracted to the more positive hydrogens.

Note that the water molecules are not aligned into a rigid, ordered structure; in practice all of the molecules are moving rapidly and randomly. Molecular dynamics simulations represent all of the intermolecular interactions with classical potentials, generating forces and acceleration via Newton's laws. Such simulations give very good descriptions of many of the properties of bulk solutions.

▶ PROBLEMS ▶

3-1.[*] Find the ratio between the strength of the gravitational force and the strength of the Coulombic force between a proton and an electron.

3-2. Energies associated with nuclear binding can be quite impressive. For example, the mass of a deuterium nucleus is $3.3435860 \times 10^{-27}$ kg; from Appendix A, the masses of the proton and neutron are $1.6726231 \times 10^{-27}$ kg and $1.6749286 \times 10^{-27}$ kg respectively. Use Einstein's formula $E = mc^2$ to calculate the energy difference between a mole of deuterium nuclei and a mole of separated protons plus neutrons.

3-3.[*] *Escape velocity* is defined as the velocity v_{esc} needed for an object on the surface of a planet or satellite to escape its gravitational pull. This means that the total energy (kinetic plus potential, where the potential energy as $r \to \infty$ vanishes, as in Equation 3.10) is at least zero.

(a) Show that the escape velocity is independent of mass.
(b) Find the escape velocity from the surface of the Earth (the mass of the Earth is 6.0×10^{24} kg; the radius is 6378 km).

3-4. An American TV comedy series of the 1980s, *WKRP in Cincinnati*, depicted life at a fictional radio station. One holiday season, the station management did a promotion that encouraged listeners to show up in a local parking lot for a surprise reward. They sent up a helicopter over the parking lot, then pushed out turkeys— reasoning that the birds would gently fly down, be caught in the lot below, and become holiday dinners. Unfortunately, turkeys cannot fly.

Assume that the turkeys were released 500 m above the ground, that the average turkey had a mass of 10 kg, and that little of the kinetic energy was dissipated on the way down. Calculate the energy released on impact. To compare this kinetic energy to chemical energy, find the amount of gasoline needed for an equivalent energy release (the explosion of gasoline produces approximately 50 kJ of energy per gram).

3-5.* Suppose we start at time $t = 0$ by extending a spring (force constant k, one moving mass m) a distance L, then releasing it ($x(0) = L$, $v(0) = 0$). Write a complete expression for the position and velocity with time. Use this to calculate the potential energy and kinetic energy, and show that while each of these quantities oscillates, the sum of the potential and kinetic energy is constant.

3-6. For two argon atoms interacting via the Lennard-Jones potential, find:

  (a) the separation which gives the minimum potential energy
  (b) the separation which gives the largest attractive force
  (c) the separation which gives the largest repulsive force

3-7.* A typical diffraction grating might have 1200 lines per mm. Suppose a 1-mm diameter argon-ion laser beam shines on the grating, in the geometry of Figure 3.11. Argon-ion lasers have two strong light components ($\lambda = 488$ nm and $\lambda = 514$ nm) so the grating will send the $n = 1$ diffracted spots in different directions, and will thus separate the two colors of light.

At a distance 1 m from the diffraction grating, what will be the separation between the spots of 488 nm and 514 nm light?

3-8. Solutions you encounter in the laboratory will always be nearly electrically neutral. The net charge on all of the cations will almost exactly balance the net charge on all of the anions. To see why this is so, imagine a solution with a slight charge imbalance: 1 mM excess positive charge, for example Na^+. Assume this solution is placed into a 1 liter, spherical container.

  (a) Using only Coulomb's law, calculate the repulsive force on a sodium ion sitting near the inside surface of the container. This force is applied on a very small area (the radius of a sodium ion is 95 pm). Recalling that pressure is defined as force per unit area, calculate the outward pressure exerted on the sodium ion. If you used SI units for everything, the pressure you calculated is in Pascals; compare this to typical atmospheric pressure (101,325 Pascals).

  (b) The potential energy of this much stored charge can be calculated to be $U = +0.15\frac{Q^2}{\pi \mathcal{E}_0 R}$, where R is the radius of the container and Q is the total charge. Calculate this energy and find the amount of gasoline needed for an equivalent energy release (as noted above, the explosion of gasoline produces approximately 50 kJ of energy per gram)

(c) The answer in part (a) above is actually somewhat of an overestimate, because the solvent itself will rearrange to partially block the field from the excess charge. The force will be scaled down by the *dielectric constant*. The dielectric constant for a typical nonpolar solvent, such as hexane, differs by about a factor of 50 from the dielectric constant for water. Which solvent has the larger dielectric constant value, and why? (The larger dielectric constant turns out to be 78.)

3-9. The diffraction equation (3.28) is a simplified version for light which comes in perpendicular to a surface. For light which hits a surface at an angle θ_{in}, the angles θ_{out} which give reinforced reflections satisfy the equation

$$n\lambda = d(\sin\theta_{in} - \sin\theta_{out})$$

Suppose you wish to determine the spectrum of a beam of light. You could use a diffraction grating to separate ("disperse") two closely-spaced wavelengths (say, 570 nm and 580 nm). Will you separate these two components more by hitting the grating at near-normal incidence ($\theta_{in} \approx 0$) or near-grazing incidence ($\theta_{in} \approx \pi/2$)?

3-10. Polonium crystals have a simple cubic structure with atoms separated by $d = 336.6$ pm. If a polonium crystal is exposed to X-rays with $\lambda = 154$ pm (the most common source wavelength), find the angle 2θ which corresponds to the deflection of the strongest scattered component.

3-11. The most common unit of pressure in the United States is pounds per square inch (psi). The pound is a unit of weight, and is the same as the force exerted by gravity on a 0.454 kg mass. One inch is equal to .0254 meters. Convert a pressure of 1 atm into psi.

3-12.* Find the pressure (force per unit area) exerted by a 1 m high column of mercury (density $= 13.6$ kg $\cdot$ L^{-1}) on the base of the column, in Pascals (the answer is independent of the width of the column). Compare this to atmospheric pressure.

3-13.* Deuterium (D), the isotope of hydrogen with one neutron and one proton, has mass 3.3445×10^{-27} kg. H, the isotope with no neutrons, has mass 1.6735×10^{-27} kg. The potential is nearly the same for H_2, HD, and D_2. Compare the vibrational frequencies of these molecules.

3-14. A 100 g ball is suspended from a spring. At time $t = 0$ the spring is neither compressed nor extended, and the velocity is 1 m $\cdot$ s^{-1} (going up). The ball goes through one complete oscillation (up and down) in one second. Calculate how far the ball is extended when the motion reaches its upper limit.

3-15.* Which is lower in energy, one mole of H_2 plus one mole of Cl_2, or two moles of HCl? How large is the energy difference?

3-16. Use the dissociation energies in Tables 3.2 and 3.4 to determine the energy produced by the complete combustion of one mole of methane (CH_4) to produce carbon dioxide and water.

3-17. Which of the diatomic molecules in Table 3.2 has the highest vibrational frequency? Which has the lowest?

Chapter 4

Introduction to Statistics
and Statistical Mechanics

> Statistical thinking will one day be as necessary for efficient citizenship as the
> ability to read and write.

H. G. Wells (1866–1946)

Statistics play an absolutely central role in chemistry, because we essentially never see one molecule decompose, or two molecules collide. In addition, the energies associated with the making and breaking of individual chemical bonds are quite small. Even a very highly exothermic reaction such as the combustion of hydrogen releases only 4×10^{-19} J of energy per hydrogen molecule. So practical processes involve very large numbers of molecules. However, when 1 g of hydrogen gas burns in oxygen to give water, 6×10^{23} hydrogen atoms undergo a fundamental change in their energy and electronic structure. The properties of the reactive mixture can only be understood in terms of averages. In fact, many of the measurable quantities commonly used in chemistry cannot be defined for a single molecule. There is no such thing as the pressure or entropy of a helium atom, and the temperature is difficult to define—yet temperature, entropy and pressure are macroscopic, measurable, averaged quantities of great importance.

You might think that the statistical nature of chemical processes is a tremendous complication. In fact, however, it often simplifies life enormously because averaged properties can be quite predictable. You can go to Las Vegas or Atlantic City, put one dollar in a slot machine, and win one million dollars; or you could go through a semester's tuition. The uncertainty in your winnings is quite large, which generates the excitement of gambling. However, casino managers know the odds in their games, and they know that, averaged over millions of visitors a year, they will come out ahead—their uncertainty is a much smaller fraction of the total bet. Similarly, at room temperature and normal pressures, about 10^{27} gas molecules hit a 1 m² window in 1 second. The pressure exerted by the gas comes from the total momentum transfer from all of these collisions. But at atmospheric pressure, the force on each side on a 1 m² window is the same as would be exerted by over 10,000 kg of mass! If the pressure on the opposite sides of the window varied by any significant fraction of this, the net force could easily break glass.

In this chapter we will discuss two statistical problems which are particularly relevant for chemistry and physics. First we will describe the distribution generated by *random walk* processes which are equally likely to be positive or negative (the ***normal*** or ***Gaussian*** distribution). We will derive this distribution under fairly stringent assumptions, but it turns out to be extremely general and useful. For example, experimental uncertainties in the laboratory usually follow this distribution, and our results thus predict how these errors are reduced by averaging. Random walks can also be used to describe the rate at which leakage from a hazardous waste site approaches an underground stream, the motion of a dust particle on the surface of a liquid, or the way odors are spread through the air.

The other problem we will discuss is the most likely distribution of a fixed amount of energy between a large number of molecules (the *Boltzmann distribution*). This distribution leads directly to the ideal gas law, predicts the temperature dependence of reaction rates, and ultimately provides the connection between molecular structure and thermodynamics. In fact, the Boltzmann distribution will appear again in every later chapter of this book.

4.1 THE "RANDOM WALK" PROBLEM

One of the most basic problems in statistics is called the *random walk problem*. Suppose you take a total of N steps along a north-south street, but before each step you flip a coin. If the coin comes up heads, you step north; if the coin comes up tails, you step south. What is the probability that you will end up M steps north of your starting point (in other words, the probability that you will get M more heads than tails)?

If you toss N coins, the total number of distinct possible outcomes is 2^N since there are two ways each coin can fall. For example, if $N = 3$ the eight outcomes are

$$\text{HHH, HHT, HTH, THH, TTH, THT, HTT, HHH}$$

The number of these outcomes which have exactly n_H heads and n_T tails is given by the ***binomial distribution***:

$$\begin{matrix} \text{Number of outcomes with} \\ n_H \text{heads and } n_T \text{tails} \end{matrix} = \frac{N!}{n_H! \, n_T!}; n_H + n_T = N \qquad (4.1)$$

The probability $P(n_H, n_T)$ of getting exactly n_H heads and n_T tails is given by dividing Equation 4.1 by the total number of possible outcomes 2^N.

$$P(n_H, n_T) = 2^{-N} \frac{N!}{n_H! n_T!} \tag{4.2}$$

Again using $N = 3$ as an example, the probability of getting two heads (and hence one tail) is

$$P(2, 1) = 2^{-3} \frac{3!}{2!1!} = \frac{1}{8} \frac{6}{2 \cdot 1} = \frac{3}{8} \tag{4.3}$$

The sum of the probabilities for all the possible values of n_H must be one, because we will always get some number of heads.

$$\sum_{n_H=0}^{n_H=N} P(n_H, N - n_H) = 1 \tag{4.4}$$

which you can also verify explicitly for $N = 3$, since $P(0, 3) = P(3, 0) = 1/8$ and $P(2, 1) = P(1, 2) = 3/8$.

Now, in order for you to end up M steps north, you need M more heads than tails:

$$n_H = n_T + M \tag{4.5}$$

which implies

$$\begin{aligned} n_H &= (N - n_H) + M \;\{\text{combining 4.1 and 4.5}\} \\ 2n_H &= N + M \\ n_H &= \frac{N + M}{2}; n_T = N - n_H = \frac{N - M}{2} \end{aligned} \tag{4.6}$$

so the solution to the problem is

$$P(M) = 2^{-N} \frac{N!}{\left(\frac{N + M}{2}\right)! \left(\frac{N - M}{2}\right)!} \tag{4.7}$$

Factorials can be calculated for $N < 100$ on hand calculators. The distribution for $N = 10$ is graphed as the dots in Figure 4.1. Since N is even, only even values of M are possible. Notice that the distribution looks very similar to the Gaussian function we first discussed in Section 2.2.

You end up where you started ($M = 0$) less than 1/4 of the time, since $P(0) = 252/1024$; however, almost 2/3 of the time (652/1024) you get $M = 2, 0$, or -2, and you are within two steps of the start.

You are equally likely to end up with more heads or more tails. Thus, if you tried this a large number of times, the expected average value of M (written $\overline{M}$) would be

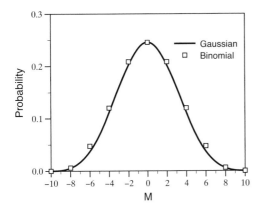

FIGURE 4.1 ▶ **Dots**: Probabilities of getting M more heads than tails out of 10 coin flips. **Solid line**: Gaussian function, as introduced in Section 2.2

zero, since the distribution is symmetric $(P(M) = P(-M))$. However, most of the time you will end up away from the start in some direction. We can quantify this by calculating the expected average value of some quantity which measures the deviation from equality and is always positive. For example, we could find the average value of $|M|$, or we could find the average value of M^2 (which is also always positive), and then take the square root at the end.

In general, the average value of any function $f(M)$ is given by the expression:

$$\overline{f(M)} = \sum_{\text{all possible } M} f(M)P(M) \tag{4.8}$$

so we have, for example,

$$|\overline{M}| = \sum_{M=-N}^{M=N} |M|\, P(M); \quad \left(\overline{M^2}\right)^{1/2} = \left\{ \sum_{M=-N}^{M=N} M^2 P(M) \right\}^{1/2} \tag{4.9}$$

Let's explicitly calculate $\left(\overline{M^2}\right)^{1/2}$ and $|\overline{M}|$:

$$\begin{aligned}
\overline{M^2} &= (-10)^2(1/1024) + (-8)^2(10/1024) + (-6)^2(45/1024) \\
&\quad + (-4)^2(120/1024) + (-2)^2(210/1024) + (0)^2(252/1024) \\
&\quad + (2)^2(210/1024) + (4)^2(120/1024) + (6)^2(45/1024) \\
&\quad + (8)^2(10/1024) + (10)^2(1/1024) \\
&= 10 \\
\left(\overline{M^2}\right)^{1/2} &= \sqrt{10} \approx 3.1
\end{aligned} \tag{4.10}$$

$$\begin{aligned}
|\overline{M}| &= |(-10)|\,(1/1024) + |(-8)|\,(10/1024) + |(-6)|\,(45/1024) \\
&\quad + |(-4)|\,(120/1024) + |(-2)|\,(210/1024) + (0)(252/1024) \\
&\quad + 2(210/1024) + (4)(120/1024) + (6)(45/1024) \\
&\quad + (8)(10/1024) + (10)(1/1024) \\
&= 2.46
\end{aligned}$$

Thus "on average" we end up about three steps from where we started, but to be more precise we have to specify just what we are averaging. It turns out that for any possible distribution, $\left(\overline{M^2}\right)^{1/2} \geq |\overline{M}|$.

4.2   THE NORMAL (GAUSSIAN) DISTRIBUTION

Equation 4.7 is very difficult to use for even a few hundred random events because factorials grow rapidly. However, the agreement between the binomial values and a Gaussian curve in Figure 4.1 is not accidental. It can be shown that if $N \gg 1$ and $M \ll N$,

$$P(M) = 2^{-N}\frac{N!}{\left(\dfrac{N+M}{2}\right)!\,\left(\dfrac{N-M}{2}\right)!} \approx Ce^{-M^2/2N} \tag{4.11}$$

which is a very much simpler formula to use. The function $e^{-M^2/2N}$ is a Gaussian function with standard deviation $\sigma = \sqrt{N}$, as discussed in Chapter 2. The distribution of probabilities in Equation 4.11 is sometimes called the *normal distribution*. Even for $N = 10$ a Gaussian function is a fairly good approximation to the binomial expression for small M; and $M \approx N$ is extremely improbable for large N anyway (see Problem 4-1).

The constant C comes from the requirement that the sum of all the probabilities equals one. If N is large, we can approximate M as a continuous variable instead of something restricted to integral values (and this will be more realistic for chemical and physical applications). Then we have

$$\int\limits_{M=-\infty}^{M=+\infty} P(M)\,dM = 1 = C\int\limits_{M+-\infty}^{M=+\infty} e^{-M^2/2N}\,dM \tag{4.12}$$

This last integral is identical to Equation 2.29, with the substitutions $x = M$ and $\sigma = \sqrt{N}$:

$$\int\limits_{M=-\infty}^{M=+\infty} e^{-M^2/2N}\,dM = \sqrt{2\pi N} \tag{4.13}$$

Substituting Equation 4.13 into 4.12 gives $C = 1/\sqrt{2\pi N}$:

$$P(M)\, dM = \frac{1}{\sqrt{2\pi N}} e^{-M^2/2N}\, dM \qquad (4.14)$$

Now we can see how this distribution behaves for large values of N. For example, if we toss 10,000 coins, we have $\sigma = \sqrt{N} = 100$ in Equation 4.14. We would expect to get, on average, 5000 heads and 5000 tails ($M = 0$), and indeed the maximum of $P(M)$ occurs at $M = 0$. But if $M \ll 100$, $P(M)$ is only slightly smaller than $P(0)$. Thus some deviations from exact equality are quite likely.

We can calculate expectation values for this continuous distribution in much the same way as we calculated them in the last section for ten coin tosses. The generalization of Equation 4.8 for a continuous distribution is:

$$\overline{f(M)} = \int_{\text{all possible } M} f(M) P(M)\, dM \qquad (4.15)$$

This can be used to show, for example, that

$$\left(\overline{M^2}\right)^{1/2} = \left(\frac{1}{\sqrt{2\pi N}} \int M^2 e^{-M^2/2N}\, dM\right)^{1/2} = \sqrt{N} \qquad (4.16)$$

which you can verify using the definite integrals in Appendix B.2.

Equation 4.16 shows that the width of the distribution is proportional to the square root of the number of steps. So $\sqrt{N}$ (which is also the standard deviation σ) provides a measure of the *fluctuations* of M from its average value of zero. After 10,000 random steps, we will be, on average, about $\sqrt{10,000} = 100$ steps away from the starting point. But if we take 1,000,000 random steps (100 times more steps) we will only end up on average 1000 steps from the start (10 times as far); out of all of these steps, in effect 999,000 (99.9%) canceled each other out.

Often we are interested in finding the probability that M is within some range. For example, we might want to know how likely we are to get more than 52% heads out of 10,000 coin tosses. In this case we need to integrate $P(M)$ over limits other than $\pm\infty$. As noted in Chapter 2, this integral is not given by any simple function, but it can be calculated by a computer. The fraction of $P(M)$ between some commonly used limits is shown in Table 4.1. For a more detailed table see Appendix B, or references [1] and [4].

When $z \gg 1$, the following approximate formula is useful:

$$\frac{1}{\sigma\sqrt{2\pi}} \int_{z\sigma}^{\infty} e^{-x^2/2\sigma^2}\, dx \approx \frac{e^{-(z^2/2)}}{z\sqrt{2\pi}} \qquad (z \gg 1) \qquad (4.17)$$

We can illustrate the application of Table 4.1 with a few examples. 95% of the area under the curve is within $\pm 1.96\sigma$ of $M = 0$. Thus we can say with 95% confidence

TABLE 4.1 ▶ The Area Under a Gaussian Curve between Different Limits

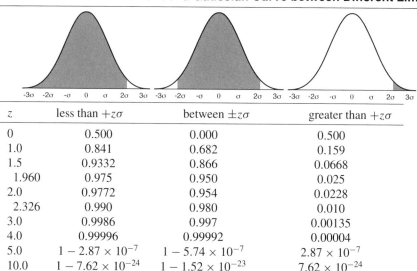

z	less than $+z\sigma$	between $\pm z\sigma$	greater than $+z\sigma$
0	0.500	0.000	0.500
1.0	0.841	0.682	0.159
1.5	0.9332	0.866	0.0668
1.960	0.975	0.950	0.025
2.0	0.9772	0.954	0.0228
2.326	0.990	0.980	0.010
3.0	0.9986	0.997	0.00135
4.0	0.99996	0.99992	0.00004
5.0	$1 - 2.87 \times 10^{-7}$	$1 - 5.74 \times 10^{-7}$	2.87×10^{-7}
10.0	$1 - 7.62 \times 10^{-24}$	$1 - 1.52 \times 10^{-23}$	7.62×10^{-24}

that M after 10,000 coin tosses will be in the range ±196, since $\sigma = \sqrt{N} = 100$. Equivalently we can say, with 95% confidence, that $n_H = (N + M)/2$ after 10,000 coin tosses will be between 4902 and 5098 (Problem 4-2).

Notice that the probability is *very* small for deviations much greater than σ. Getting more than 52% heads out of 10,000 tosses requires $n_H \geq 5200$, or $M = n_H - n_t \geq 400$; this is the area farther than $+4\sigma$, or 3.91×10^{-5}. The probability of getting more than 55% heads out of 10,000 tosses ($M \geq 1000$) is the area past 10σ, or less than one part in 10^{23}.

In general, as we increase the number of coin tosses, the *absolute* expected deviation from exactly $N/2$ heads $\left(\left(\overline{M^2} \right)^{1/2} = \sqrt{N} \right)$ grows, but the *fractional* expected deviation from exactly 50% heads $\left(\left(\overline{M^2} \right)^{1/2} /N = 1/\sqrt{N} \right)$ shrinks. For example, the probability of getting more than 50.2% heads out of 1,000,000 tosses ($M \geq 4000$) is the same as the probability of getting more than 52% heads after 10,000 coin tosses ($M \geq 400$). (Problem 4-3.)

4.3 APPLICATIONS OF THE NORMAL DISTRIBUTION IN CHEMISTRY AND PHYSICS

The most important results of the last section can be summarized as follows:

1. The likely fluctuation from the most probable result is proportional to the square root of the number of random events ($\sqrt{N}$).

2. The probability of observing a fluctuation which is much larger than the standard deviation σ is extremely small.

3. The fractional fluctuation from the most probable result scales as the inverse square root of the number of random events $(1/\sqrt{N})$.

These results turn out to be applicable to much more than just coin toss problems. A few examples are given below.

4.3.1 Molecular Diffusion

Suppose a chunk of dry ice evaporates at the center of a long tube. The gaseous carbon dioxide molecules initially travel to the left or right with equal probability. Let's over-simplify the problem a bit to begin with: assume that, at one specific time, each CO_2 molecule moves at a speed s left or right (typically ≈ 400 m $\cdot$ s^{-1}), and each molecule travels a distance λ (typically ≈ 10 nm) before it collides with another gas molecule. Each collision completely randomizes the velocity.

Under these assumptions, the probability of finding a carbon dioxide molecule $M\lambda$ from its starting point after N collisions is *mathematically exactly the same* as the coin toss probability of M more heads than tails. The root-mean squared distance traveled from the starting point will be $\left(\overline{M^2}\right)^{1/2} \lambda = \sqrt{N}\lambda$—proportional to the square root of the number of steps, or equivalently proportional to the *square root* of the travel time. If there were no collisions, the mean distance traveled would be proportional to the first power of the travel time $(x = v_x t)$.

Of course, real molecules have a distribution of velocities, and they do not all travel the same distance between collisions. However, there is a remarkable theorem which is proven in advanced physics courses which shows that *essentially every purely random process gives a normal distribution*:

$$P(M)\,dM = \frac{1}{\sigma\sqrt{2\pi}}e^{-M^2/2\sigma^2}\,dM \tag{4.18}$$

where now we use σ, which simply reflects the width of the distribution, instead of the unknown number of random events N. So the results for coin tosses turn out to be extremely general. If all of the molecules start at time $t = 0$ at the position $x = 0$, the concentration distribution of $C(x, t)$ at later times is a Gaussian:

$$C(x, t) \propto \exp\left(-\frac{x^2}{4Dt}\right) \tag{4.19}$$

where D is called the **diffusion constant**. Comparing 4.19 to the standard form of the Gaussian (Equation 4.18) shows that the standard deviation σ of this Gaussian is $\sigma = \sqrt{2Dt}$, so we immediately write:

$$\left(\overline{x^2}\right)^{1/2} \text{ (RMS distance traveled from start)} = \sigma = \sqrt{2Dt} \tag{4.20}$$

TABLE 4.2 ▶ **Diffusion Constants for some Gases and Liquids**

Molecule	Conditions	$D(\times 10^{-5}\ m^2 \cdot s^{-1})$
Hydrogen	in hydrogen at STP	15
	in air at STP	7.0
Nitrogen	in air at STP	1.85
Carbon dioxide	in air at STP	1.39
	in CO_2 at STP	1.0
Iodine	dissolved in liquid hexane, 25C	.0004
Water	in water at 25C	.00024
Hemoglobin	in water at 25C	.000007

Some typical values for gases and liquids are given in Table 4.2.

Suppose we fill a beaker with carbon dioxide gas, perhaps from dry ice evaporating in the bottom, and then remove the dry ice. Does the carbon dioxide instantly dissipate into the atmosphere? *No*; Equation 4.20 shows that in one second, the average CO_2 molecule will only migrate

$$\sqrt{2 \cdot (1.29 \times 10^{-5}) \cdot 1}\ m = .0053\ m = 5.3\ mm.$$

After 100 seconds, the average migration is only 53 mm. So the beaker will remain filled with CO_2 for an extended period. A common lecture demonstration involves "pouring the gas" from such a beaker over a candle flame, which extinguishes the candle. The carbon dioxide gas can be poured through air because it is heavier than air, yet the low rate of diffusion guarantees that the CO_2 molecules will mainly remain together.

4.3.2 Error Bars

The single most important application of statistical methods in science is the determination and propagation of *experimental uncertainties*. Quantitative experimental results are never perfectly reproducible. Common sources of error include apparatus imperfections, judgments involved in laboratory technique, and innumerable small fluctuations in the environment. Does the slight breeze in the lab affect a balance? When a motor starts in the next building, does the slight power surge affect a voltmeter? Was the calibrated volumetric flask perfectly clean?

Sometimes we know a great deal about the expected statistics of a measurement. Suppose we actually flip the same coin 10,000 times, and get 5500 heads; we showed that the chance of getting this many heads or more is less than 10^{-23}. We could conclude that we were just *extremely* lucky. However, it is more reasonable to conclude that something is biased about the coin itself or the way we tossed it, so that heads and tails do not really have equal probability. Would we draw the same conclusion if we got 5200 heads (the chance of getting this many heads or more is 3.91×10^{-5})? Would we draw the same conclusion if we got 5050 heads?

We showed in the last section that there is a 95% chance that after 10,000 coin tosses, the number of heads would be between 4902 and 5098. The most common definition treats error bars as such a **95% *confidence limit***. Table 4.1 shows that 95% of the area in a Gaussian is within $\pm 1.96\sigma$ from the center, so the error bars are $\pm 1.96\sigma$ and the number of heads n_H after 10,000 coin tosses is "5000 ± 98." For the reasons discussed below, we would probably round it to 5000 ± 100. So we would not be sufficiently surprised by 5050 heads to judge that something was wrong; we would be surprised enough by 5200 heads.

There is nothing magical about the choice of 95% in the confidence limits; it is merely a common compromise between setting error bars so wide that you never draw any conclusions, and setting them so narrow that you are led to false conclusions. In some cases higher or lower values are appropriate. For example, 99.99% confidence limit error bars would be about $\pm 4\sigma$ from Table 4.1.

Sometimes a measurement involves a single piece of calibrated equipment with a known measurement uncertainty value σ, and then confidence limits can be calculated just as with the coin tosses. Usually, however, we do not know σ in advance; it needs to be determined from the spread in the measurements themselves. For example, suppose we made 1000 measurements of some observable, such as the salt concentration C in a series of bottles labeled 100 mM NaCl. Further, let us assume that the deviations are all due to random errors in the preparation process. The distribution of all of the measurements (a histogram) would then look much like a Gaussian, centered around the ideal value. Figure 4.2 shows a realistic simulated data set. Note that with this many data points, the near-Gaussian nature of the distribution is apparent to the eye.

To find the width of the distribution, we evaluate the measured average (mean) concentration and the root-mean-squared deviation from the average:

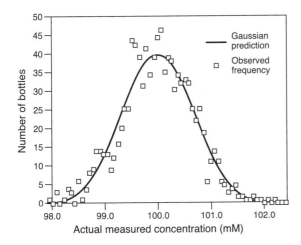

FIGURE 4.2 ▶ Simulated histogram of 1000 different random measurements, compared with a Gaussian distribution.

$$\text{mean concentration} \ = \ \overline{C} = \sum_{i=1}^{N} \frac{C_i}{N}$$

$$\text{r.m.s. deviation from average} \ = \ \sigma = \left\{ \sum_{i=1}^{N} \frac{(C_i - \overline{C})^2}{N} \right\}^{1/2} \tag{4.21}$$

For this particular data set, $\overline{C}$ turns out to be 99.944 mM, but one glance at Figure 4.2 shows that not all of those digits are significant. σ turns out to be 0.66 mM, so we would *calculate* that the concentration of a typical bottle is 99.944±(1.96·0.66) mM = 99.944 ± 1.29 mM. In practice, we would not *report* so many digits: the usual convention is that the *last reported digit should be uncertain by an amount between 3 and 30 units.* So we would write 99.9 ± 1.3 mM (the last digit, in the tenth-millimolar position, is uncertain by 13 units) in writing confidence limits *for the distribution.* This implies that 95% of the time a bottle selected at random would have a concentration between 98.6 and 101.2 mM.

We can also generalize result (3) at the beginning of Section 4.3 to say that *the average of N measurements is expected to be in error by an amount which is proportional to $1/\sqrt{N}$.* This is the principle behind **signal averaging.** The average of 1000 trials is expected to be $\sqrt{1000}$ times more accurate than the result of a single trial. So we would report 99.974 ± (1.96 · 0.66)/$\sqrt{1000}$ mM = 99.94 ± 0.04 mM (again to the correct number of significant digits) in writing confidence limits *for the mean.*

Note now that the labeled mean concentration is outside of the error bars, which as we noted before were based on 95% confidence limits. This means that, if the mean concentration were really 100 mM, there would be less than one chance in 20 that 1000 bottles, chosen at random, would give a deviation from the average which was this large. Based on these statements, we can conclude (again at the 95% confidence level) that the actual mean concentration, which we could approach in principle by measuring an extremely large number of bottles, is less than 100.00 mM.

Which is more important: confidence limits for the *mean* or confidence limits for the *distribution?* It depends on the application. The vendor of the standardized solutions above should report confidence limits for the distribution to its customers, who will use one bottle at a time (for example, in preparing a saline solution to be injected into a patient). In other cases, however, the error is in the measurement process itself. We believe that all electrons have the same mass, but 1000 measurements of electron mass will likely all give slightly different answers. Then we want to know confidence limits for the mean. In addition, 95% confidence limits for the mean are used by pollsters to predict the results of an election. The fact that individual preferences vary is not interesting; what *is* interesting is whether, on average, more than 50% of the voters prefer one specific candidate.

The approach described above only gives correct confidence for a very large number of observations, say $N > 50$. It is possible to generalize these formulas to assign statistically valid error bars for smaller numbers of observations, but the formulas are more complex (Problem 4-6 gives one example).

Unfortunately, the $1/\sqrt{N}$ factor ultimately overwhelms the patience of the experimenter. Suppose a single measurement (for example, determination of the endpoint of a titration) takes ten minutes. The average of four measurements is expected to be twice as accurate, and would only take thirty extra minutes. The next factor of two improvement (to four times the original accuracy) requires a total of 16 measurements, or another 120 minutes of work; the next factor of two requires an additional 480 minutes of work. In addition, this improvement only works for random errors, which are as likely to be positive as negative and are expected to be different on each measurement. *Systematic errors* (such as using an incorrectly calibrated burette to measure a volume) are not improved by averaging. Even if you do the same measurement many times, the error will always have the same sign, so no cancellation occurs.

Before we leave the subject of confidence limits, a few warnings are in order.

1. *Even the best statistical methods do not prevent disagreements.* For example, the choice of 95% confidence limits is arbitrary. You might want to use higher confidence limits if, for example, you were trying to determine if your control rods would absorb enough neutrons to prevent the reactor in a nuclear power plant from going critical. Thus two people can look at the same data and draw different conclusions. Notice, however, that 99.99% confidence limits are only about twice as wide as 95% confidence limits if σ is known, and if systematic errors can be ignored.

2. *Nearly impossible things happen every day.* If your chance of winning a lottery with a single ticket purchase is 10^{-7}, you can say, with 99.99999% confidence, that the ticket in your hand will not be a big winner. Yet somebody will beat these odds, or eventually nobody would buy lottery tickets.

   Suppose you draw three cards from a standard deck of 52: the three of spades, the six of diamonds, and the eight of diamonds. Should you be astonished? After all, your chance of getting these three cards in this order is only $(1/52) \cdot (1/51) \cdot (1/50)$, or less than 10^{-5}; you could say, with 99.999% confidence, that this combination should not have happened. But you had to get *some* combination of three cards, and they all have the same probability.

   A less obvious (but far more common) abuse of statistics is their use to analyze health risks. For example, we know with high accuracy the average incidence of any of hundreds of different subtypes of cancer, based on reporting by doctors over the last several decades. Suppose I select 100 towns at random, analyze the incidence of 100 different types of cancer in each of these towns over a decade, and compare these incidences to the known averages with 95% confidence limits. Out of these 10,000 combinations, on average *500* will be outside the limits and will be "statistically significant"! About 250 combinations of one town and one disease will be "statistically high," and will terrify the local population when the results are published; the 250 combinations of one town and one disease which are "statistically low" will probably be ignored.

All of the seeming paradoxes above can be resolved by remembering the difference between prediction and postdiction. You should be astonished if someone tells you in advance that you will pick the three of spades, the six of diamonds, and the eight of diamonds. You should be concerned if a town with a factory which produces a known liver carcinogen shows a statistically significant incidence of increased liver cancer. And if anyone can tell you for sure that the lottery ticket you are about to purchase will be a winner, by all means do it.

4.3.3 Propagation of Errors

Often we need to combine several laboratory measurements or do some additional data manipulation to extract specific quantities of interest. Suppose, for example, that you wish to measure the solubility product of silver chloride, which we used in our illustration of quadratic equations in Chapter 1. Consider the following procedure:

- Start with 1000.0 mL of pure water (measured with a volumetric flask to ± 0.5 mL accuracy). Add 10.000 mg silver chloride (weighed to ± 0.01 mg accuracy), then stir to produce a saturated solution.

- Separate the remaining solid from the liquid and determine the mass of the solid after it has dried, in order to determine the number of grams which dissolved. If the temperature is $10°C$, the remaining mass would be 9.112 mg, also weighed to ± 0.01 mg accuracy.

- Divide the dissolved mass by the sum of the atomic weights of silver ($107.8682 \pm .0003$ g $\cdot$ mol^{-1}) and chlorine ($35.4527 \pm .0003$ g $\cdot$ mol^{-1}), to determine the number of dissolved moles of silver chloride. This will directly give the (equal) concentration of silver ions and chloride ions. The square of the number of dissolved moles of silver chloride is the solubility product.

The procedure outlined above has many possible sources of both random and systematic error. The measurements of volume and of mass will not be perfectly accurate; if the equipment has been correctly calibrated and the laboratory technique is good, these errors are random (equally likely to be positive or negative). The masses of the two nuclei are not perfectly known, but these errors can be assumed to be random as well (the error bars are the results of many careful measurements). The silver chloride and water will both have impurities, which will tend to make systematic errors. Some impurities (e.g., chloride ions in the water) would tend to make the measured solubility product smaller than the true value. Some impurities (e.g., sodium chloride in the silver chloride) would tend to make the measured solubility product larger than the true value. Finally, even without impurities, there is one (probably small) systematic error in the procedure outlined above which will tend to make the measured solubility product larger than the true value; can you spot it?

In good laboratory procedure, the systematic errors are significantly smaller than the random errors. Assuming this is the case, the random errors are propagated as follows:

- **Addition or subtraction**: given two quantities with random errors $A \pm (\Delta A)$ and $B \pm (\Delta B)$, the sum $A + B$ or the difference $A - B$ has random error $\sqrt{(\Delta A)^2 + (\Delta B)^2}$.

  This formula is easy to verify with the "coin toss" distribution. As discussed above, with 10,000 coin tosses you have 5000 ± 98 heads. You can also verify (by calculating σ) that with 2500 coin tosses you have 1250 ± 49 heads, but for 12,500 tosses you get "6250 ± 109.56733" heads, not 6250 ± 147 heads as you would get by just adding the error bars for 10,000 tosses and 2500 tosses. We would report 6250 ± 110 heads, using the rounding off convention discussed above (last digit uncertain between 3 and 30 units).

  In propagating errors it is generally advisable to keep one or two extra digits in intermediate results, and round off only when you get to the final result. Applying this rule to the solubility product measurement, the mass of dissolved silver chloride is 0.888 ± 0.0142 mg, and the formula weight of silver chloride is 143.3279 ± 0.00042 g $\cdot$ mol^{-1}.

- **Multiplication or division**: given two quantities with random errors $A \pm (\Delta A)$ and $B \pm (\Delta B)$, the product or quotient $(C = A \cdot B$ or $C = A/B)$ has a random error given by $(\Delta C)/C = \sqrt{((\Delta A/A)^2 + ((\Delta B)/B)^2}$

Notice that it is the *fractional* error which counts in multiplication or division. For example, if $A = 100 \pm 3$ (3% error) and $B = 20.0 \pm 0.8$ (4% error), the product AB will have a 5% error using this rule above ($AB = 2000 \pm 100$). The number of moles of dissolved silver chloride is calculated to be

$$\frac{0.000888 \pm 0.0000142 \text{ g}}{143.3279 \pm 0.00042 \text{ g} \cdot \text{mol}^{-1}} = (6.1956 \times 10^{-6}) \pm (9.9074 \times 10^{-8}),$$

which we would round to $(6.20 \pm 0.10) \times 10^{-6}$ moles. The volume is $1000 \pm .05$ mL, so the concentration is $(6.20 \pm 0.10) \times 10^{-6}$ M. The error in the volume measurement makes no discernible difference in the error bars.

Often the data analysis requires multiplication by some number B such as *Avogadro's number*, which is typically known to much higher accuracy than any of the measured data. In that case, the rule for multiplication above simplifies immediately to $\Delta C/C = \Delta A/A$—the fractional error after multiplication by a constant is unchanged.

- **Raising to a power**: given a quantity with random error $A \pm (\Delta A)$, the quantity $C = A^n$ has a random error given by $(\Delta C)/C = n(\Delta A)/A$

Notice that the fractional error in A^2 is twice the fractional error in A, not $\sqrt{2}$ times the fractional error as would happen if you multiplied two independent numbers with the same error. The solubility product is the square of the concentration, so it is $(3.83 \pm 0.12) \times 10^{-11}$ M^2. By the way, notice that each individual measurement was made with much higher accuracy than the final result—correct error propagation is quite important.

4.4   THE BOLTZMANN DISTRIBUTION

A different distribution arises if the random process is subject to some *constraints*. For example, we might know the average kinetic energy of the gas molecules in a bulb, and we might like to predict how the energy is distributed among the molecules, subject to the knowledge that only distributions which give our known average kinetic energy are possible.

The applications are very general, but we will again illustrate the distribution with a "coin toss" problem. A chemistry instructor, eager to improve his course ratings, decides to end his last lecture in dramatic fashion. He spends his monthly salary to buy 20 gold coins, which he tosses into a bag along with nine chocolate coins. The 10 students come up, one at a time, to draw coins from the bag just before filling out the evaluation forms. Each student can keep any gold coins he or she draws, but must stop when a chocolate coin is picked. The gold coins left after the ninth chocolate coin is picked go to the tenth student. What is the most likely distribution?

Figure 4.3 shows a few of the many different ways the 20 gold coins can go to 10 students. Before you go further, estimate the relative likelihood of these distributions.

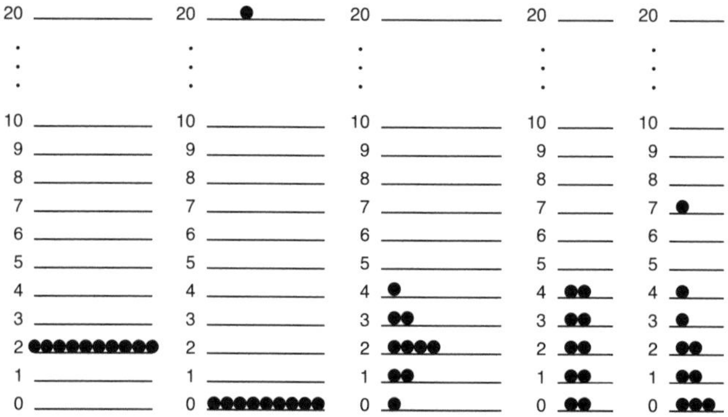

FIGURE 4.3 ▶ Five of the many possible ways twenty coins can be distributed among ten students. Each student is represented by a circle. Thus, for example, the column on the left illustrates the case where each student gets two coins.

Assume n_0 students end up with no coins, n_1 students with one coin, and so forth. Then the total number of students N (10 in this case) is:

$$n_0 + n_1 + n_2 + \cdots \left(= \sum_{j=0}^{\infty} n_j \right) = 10 \tag{4.22}$$

The total number of coins C (20 in this case) serves as the constraint. The number of ways each distribution can be generated, which we will call Ω, turns out to be given by what is called the **multinomial expansion**:

$$\Omega = \frac{N!}{n_0! \, n_1! \, n_2!} \cdots \tag{4.23}$$

Consider, for example, the second distribution. One student gets all of the coins ($n_0 = 9$, $n_{20} = 1$, and all other $n_i = 0$), giving $\Omega = 10$ in Equation 4.23. This makes sense because the "big winner" could be any one of the ten students, so there are 10 different possibilities for this distribution. On the other hand, the first distribution, which treats everyone exactly the same, gives $\Omega = 1$ in Equation 4.23 ($n_2 = 10$, all other $n_i = 0$). Neither distribution is likely.

The probability gets higher if more groups of students are included in the distribution. For example, the third distribution, which has its maximum at the expected average with a range of values, gives $\Omega = 37{,}800$. However, it is far from the most probable distribution. The fourth distribution gives $\Omega = 113{,}400$; the fifth distribution, which is more probable than any other, gives $\Omega = 151{,}200$. Notice that the most probable distribution is *biased towards the smallest values*. In this case, 30% of the students get no coins, and are really angry when they write their course evaluation forms; another 20% got only one coin, or less than their expected "fair share." So 50% get less than the average, but only 30% get more than average. (The reader may decide whether this experiment produced the intended result of uniformly glowing course evaluations).

Just as with the binomial distribution, calculating factorials is tedious for large N. The binomial distribution converged to a Gaussian for large N (Equation 4.11). The most probable distribution for the multinomial expansion converges to an exponential:

$$n_j = n_0 \exp(-\alpha j) \quad (N, C \gg 1) \tag{4.24}$$

where α is some constant. Notice this implies that the ratio of the populations of adjacent levels is fixed by α:

$$\frac{n_1}{n_0} = \frac{n_2}{n_1} = \frac{n_3}{n_2} = \frac{n_4}{n_3} = \exp(-\alpha) \tag{4.25}$$

The constant $\exp(-\alpha)$ gets larger as the average number of coins C/N increases, but is always less than one. This distribution for the case $C/N = 2$ (two coins on average per student) and a large number of students is graphed in Figure 4.4.

Now let's translate this "coin toss" problem back into a chemical application. Suppose molecules only had energy levels which were multiples of some specific packet, or "quantum." Figure 4.3 would also give the different distributions for 20 "quanta" among 10 molecules; Figure 4.4 would give the expected distribution for a large number of molecules if there were an average of two quanta of energy per molecule.

More generally, of course, the possible energy levels are not equally spaced, and they might even have a continuous distribution. However, the distribution of energy between the different available states of a molecule (kinetic energy of translation, vibrational energy, and so forth) still follows *exactly this same pattern*. The generalization of Equation 4.25 is that the most probable distribution of populations between two states i and i, with energies E_i and E_j respectively, is given by:

$$\frac{n_i}{n_j} = e^{-\beta(E_i - E_j)} \qquad (4.26)$$

The quantity $\beta(E_i - E_j)$ plays the same role in Equation 4.26 as α played in Equation 4.25 (in the "coin toss" problem, the "energy difference" between adjacent levels was one coin). Increasing β increases the fractional population of highly excited states, and thus increases the total energy of the system. Equation 4.26 is called the **Boltzmann distribution**, after Ludwig Boltzmann, a famous theoretical physicist.

Suppose we have a system with a large number of available states, such as a mole of gas in a box. The positions and velocities of the 6×10^{23} gas molecules are constantly changing due to collisions with the walls and collisions with each other. If we leave the gas undisturbed for a long enough time, the system will evolve to a state, commonly

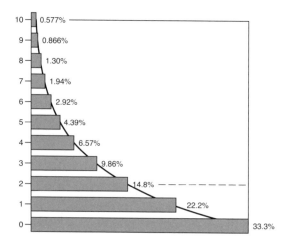

FIGURE 4.4 ▶ The most probable distribution of $2N$ coins among N students, where N is very large, is given by Equations 4.24 and 4.25. Note that only about one in seven students gets the expected "average result" (2 coins), and that most students have less than the average. Energy divides up between the available states in molecules in exactly this same way.

called *equilibrium*, where none of the macroscopic observables of the system (for example, pressure and total energy) are changing. In this case, because the number of molecules is so large, fluctuations about the most probable distribution are very small (see Section 4.3). The relative number of gas molecules with different amounts of energy (kinetic energy, internal vibrational energy, rotational energy, electronic excitation, and any other possible energies the molecules can have) will be given by Equation 4.26, with the value of β determined by the amount of energy available to the gas.

In order to make the expression dimensionless in Equation 4.26, the units of β must be (joules)$^{-1}$, or some other inverse energy. It is more convenient to rewrite β as:

$$\beta = \frac{1}{k_B T} \tag{4.27}$$

where T is called the ***temperature***, and $k_B = 1.38 \times 10^{-23}$ J $\cdot$ K^{-1} is called *Boltzmann's constant*. We can combine Equations 4.26 and 4.27 to write:

$$\frac{n_i}{n_j} = e^{-(E_i - E_j)/k_B T} \tag{4.28}$$

Equation 4.28 is the *definition* of temperature. Temperature is intrinsically a very different quantity from, for example, pressure or volume. It is a *macroscopic* and *statistical* concept. It makes no sense to talk about the temperature of a single helium atom, although the energy of such an atom is well defined. But temperature is very important, because when two systems mix, energy is exchanged until the value of T (or β) is the same—even if this means the average energy is different between the two systems. For example, a balloon filled with 50% helium and 50% nitrogen will have the same temperature for the helium atoms as it will for the nitrogen molecules. However, as we will show in Chapter 8, the nitrogen molecules will be more energetic on average because of molecular vibrations and rotations.

Equation 4.28 maximizes Ω, the number of ways the energy can be distributed consistent with the known total energy. This led to the concept of ***entropy***,

$$S = k_B \ln \Omega \tag{4.29}$$

Entropy is also a macroscopic and statistical concept, but is extremely important in understanding chemical reactions. It is written in stone (literally; it is the inscription on Boltzmann's tombstone) as the equation connecting thermodynamics and statistics. It quantifies the second law of thermodynamics, which really just asserts that systems try to maximize S. Equation 4.29 implies this is equivalent to saying that they maximize Ω, hence systems at equilibrium satisfy the Boltzmann distribution.

4.5 APPLICATIONS OF THE BOLTZMANN DISTRIBUTION

4.5.1 Distribution of Gases Above the Ground

Recall from Chapter 3 that the potential energy due to gravity is $U = mgh$ near the Earth's surface. The acceleration due to gravity is $g = 9.8$ m $\cdot$ s^{-2} independent of mass; the Earth pulls proportionately just as hard on an oxygen molecule ($m = 5.3 \times 10^{-26}$ kg) as it does on a bowling ball ($m = 7$ kg). Bowling balls sit on the ground. But we can breathe, so oxygen molecules are present above the ground. Why is this?

The number of molecules or bowling balls at height h, which we will call n_h, can be compared to the number n_0 at $h = 0$ by Equation 4.28:

$$\frac{n_h}{n_0} = e^{-(E_h - E_0)/k_B T} = e^{-mgh/k_B T}$$

At room temperature (300K) we have:

$$\frac{mg}{k_B T} = 1.26 \times 10^{-4} \text{ m}^{-1} \text{ (oxygen molecules)}$$

$$\frac{mg}{k_B T} = 1.66 \times 10^{22} \text{ m}^{-1} \text{ (bowling balls)}$$

This predicts that bowling balls do not float any significant distance above the floor—in accord with our expectation. However, the situation for oxygen molecules is quite different. At a height of approximately 8000 m, $mgh/kT \approx 1$, and the concentration of oxygen molecules has fallen to about $1/e$ of its value at sea level (if we assume no temperature variations).

4.5.2 Velocity Distribution and Average Energy of Gases

We can use the Boltzmann distribution to give the velocity distribution for a gas at equilibrium. If we are only interested in the y-direction (in other words, we want to know the probability of finding different values of v_y, independent of the values of v_x or v_z) the Boltzmann distribution gives:

$$\frac{N(v_y)}{N(v_y = 0)} = e^{-mv_y^2/(2k_B T)} \tag{4.30}$$

This is called the ***one-dimensional velocity distribution***, since only the y direction is included. It can also be converted to a probability distribution $P(v_y)\,dv_y$, which should be interpreted as the chance of finding any one molecule with a velocity between v_y and $v_y + dv_y$:

$$P(v_y)\,dv_y = \left\{ \sqrt{\frac{m}{2\pi kT}} \right\} e^{-mv_y^2/(2k_B T)}\,dv_y \tag{4.31}$$

The term in {brackets} in Equation 4.31 is the normalization constant, chosen so that $\int P(v)\,dv = 1$.

Equation 4.31 is a Gaussian in v_y with standard deviation $\sigma = \sqrt{k_B T / m}$. The distribution is peaked at $v_y = 0$ (remember v_y can be positive or negative). The average value $\overline{v_y} = 0$, since as many molecules are going left as right. The mean-squared velocity in the y direction $\overline{v_y^2}$ (square *before* averaging) is evaluated in the same way as $\overline{M^2}$ was for the coin toss distribution in Section 4.2: this quantity is just equal to σ^2 for a Gaussian. We thus have

$$\overline{v_y^2} = \frac{k_B T}{m}; \quad \left(\overline{v_y^2}\right)^{1/2} = \sqrt{\frac{k_B T}{m}} \tag{4.32}$$

Recall here that m is the mass of a single particle; for He, $m = 6.65 \times 10^{-27}$ kg.

In general, we expect the velocity distributions in the x-, y-, and z-directions to be the same, and thus we can relate the pressure to the mean-squared speed $\overline{s^2}$:

$$\overline{v_x^2} = \overline{v_y^2} = \overline{v_z^2}; \; \overline{s^2} = \overline{v_x^2} + \overline{v_y^2} + \overline{v_z^2} = 3\overline{v_y^2} \tag{4.33}$$

hence

$$\overline{s^2} = \frac{3 k_B T}{m}; \quad \left(\overline{s^2}\right)^{1/2} = \sqrt{\frac{3 k_B T}{m}} \tag{4.34}$$

If the speed of the ith particle is s_i, and there are a total of N particles, the total energy is given by:

$$E = \sum_{i=1}^{N} \frac{m s_i^2}{2} = N\left(\frac{m\overline{s^2}}{2}\right) = 3N\left(\frac{m\overline{v_y^2}}{2}\right) \tag{4.35}$$

Substituting Equation 4.32 into Equation 4.35 gives

$$E = \frac{3}{2} N k_B T \tag{4.36}$$

This simple equation explains why there is such a thing as a lowest possible temperature (absolute zero). At that temperature, the kinetic energy would equal zero, and the molecules would be completely motionless.

As we show in Chapter 7, R in the ideal gas equation and the Boltzmann constant k_B are related by the expression

$$R = N_{\text{Avogadro}} k_B = (6.022 \times 10^{23}) \quad k_B = 8.3144 \; \text{J/(mole} \cdot K) \tag{4.37}$$

Thus we could also write

$$E = \frac{3}{2} n R T \tag{4.38}$$

where n is the number of moles. This expression only works for a monatomic gas, because we have ignored any internal energy.

The total energy of a system is a difficult quantity to measure directly. It is much easier to measure energy changes dE/dT—for example, the number of joules necessary to raise the temperature of one mole of gas by one degree Kelvin. If the gas is kept in a constant volume container, this is called the ***constant-volume molar heat capacity*** c_v, and equals $3R/2$ (independent of temperature) for a monatomic gas. Each possible direction of motion (x, y, or z) contributes $RT/2$ to the total energy per mole, or $R/2$ to the heat capacity.

4.6 APPLICATIONS OF STATISTICS TO KINETICS AND THERMODYNAMICS

Kinetic properties (rates of chemical reactions) and thermodynamic properties (equilibrium constants, energy, entropy) are described by a large number of different mathematical relations, which are usually just presented for the student to memorize. Part of the reason for this is the complexity associated with a full treatment of these properties; these subjects are taught in graduate chemistry and physics courses at every major university, and multivariate calculus is needed to formulate a rigorous treatment. Unfortunately, simple memorization does not provide much intuition.

This section takes a different approach. We will show in this section that it is possible to *rationalize* the fundamental thermodynamics and kinetic equations presented in introductory chemistry courses using only the statistical concepts we have outlined in this chapter. The goal here is to show why these equations are reasonable, not to give rigorous proofs.

4.6.1 Reaction Rates: The Arrhenius Equation

The rates of chemical reactions are generally temperature dependent, and most chemical reactions go more quickly as the temperature increases. For example, the molecule C_3H_6 (called cyclopropane) has the three carbon atoms arranged in a triangle of carbon-carbon single bonds. The $60°$ bond angle is smaller than normal for carbon, so this molecule is strained. It is less stable than the molecule propene, which has the same formula but a double bond instead of the ring. As a result, cyclopropane can spontaneously convert to propene, or *isomerize*. At high pressures the isomerization rate is proportional to the concentration of cyclopropane:

$$\frac{d[\text{cyclopropane}]}{dt} = -k_f[\text{cyclopropane}] \tag{4.39}$$

At 773K, $k_f = 5.5 \times 10^{-4}$ sec^{-1}; at 1000K the reaction is far faster ($k_f = 8.1$ sec^{-1}). Experiments show that over a wide range of temperatures we can write:

$$k_f = A \exp\left(-\frac{B}{T}\right) \tag{4.40}$$

where A and B are constants which depend on the specific reaction. The reverse reaction is also possible, but slower. If we start with pure propene, it is possible to produce cyclopropane:

$$\frac{d[\text{cyclopropane}]}{dt} = k_r[\text{propene}] \tag{4.41}$$

At 773K, $k_r = 1.5 \times 10^{-5}$ sec^{-1}.

We can understand Equation 4.40 and give a physical interpretation of A and B by viewing this decomposition as proceeding through an intermediate state called the **transition state** (Figure 4.5). In the transition state, the potential energy is higher than in cyclopropane itself, because bonds have to break for the molecule to rearrange. The energy difference between the reactant and the transition state is called the **activation energy** E_a. Of course, the atoms later regain this energy (and more) as the double bond forms in the product.

Only molecules with energy comparable to the transition state can get over this potential energy barrier and rearrange. The cyclopropane molecules have a distribution of internal energies, enforced by collisions. At any given time, a small fraction of the cyclopropane molecules will be in highly excited states with sufficient internal energy to "cross over" the barrier. The number of cyclopropane molecules with internal energy equal to the activation energy E_a is given by the Boltzmann distribution as:

$$n_{E_a} = n_0 \exp\left(-\frac{E_a}{k_B T}\right) \tag{4.42}$$

This explains the exponential term in Equation 4.41; the constant B is just the activation energy:

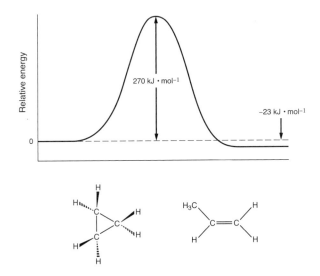

FIGURE 4.5 ▶ Relative energies of cyclopropane, propene, and the transition state between the two.

$$k_f = A \exp\left(-\frac{E_a}{k_B T}\right) \qquad (4.43)$$

Equation 4.43 is called the ***Arrhenius equation***, after Svante Arrhenius who proposed it in 1889.

Not every molecule with this much energy will rearrange. The energy has to be deposited in the correct bonds, so dissociation usually does not occur instantaneously after a collision produces a cyclopropane molecule with enough energy. If it takes too long for energy to migrate to the correct bonds, another collision may well deactivate the molecule. Thus the A factor (called the preexponential factor) depends on collision rate (assuming that each collision randomizes the internal energy, allowing more molecules to reach a high enough internal energy to dissociate) and on geometrical factors within the molecule as well. The collision frequency increases with temperature, but very slowly compared to the factor $\exp(-E_a/k_B T)$ so it is usually a good approximation to say that A is temperature independent.

4.6.2 Equilibrium Constants: Relation to Energy and Entropy Changes

Both cyclopropane and propene are stable at room temperature for extended periods. At elevated temperatures, however, the forward and backward reaction rates ultimately produce an equilibrium between cyclopropane and propene. This equilibrium is characterized by an ***equilibrium constant***

$$K = \frac{[\text{propene}]}{[\text{cyclopropane}]},$$

which is the ratio of the two concentrations. By definition, at equilibrium the concentrations remain constant, so the rate of destruction of cyclopropane molecules ($k_f[\text{cyclopropane}]$) equals the rate of creation of cyclopropane from propene ($k_r[\text{propene}]$).

$$\text{Equilibrium: } k_f[\text{cyclopropane}] = k_r[\text{propene}] \qquad (4.44)$$

which can be combined with Equation 4.43 to give:

$$
\begin{aligned}
K &= \frac{[\text{propene}]}{[\text{cyclopropane}]} = \frac{k_f}{k_r} \\
&= \frac{A_f \exp\left(-(E_{\text{transition state}} - E_{\text{cyclopropane}})/k_B T\right)}{A_r \exp\left(-(E_{\text{transition state}} - E_{\text{propene}})/k_B T\right)}
\end{aligned}
\qquad (4.45)
$$

where we have explicitly written the activation energy of the forward reaction as the energy of the transition state minus the energy of cyclopropane; the activation energy of the reverse reaction is the energy of the transition state minus the energy of propene.

As noted above, not every molecule with an energy equal to the activation energy reacts; the energy has to be in the correct bonds to form the transition state. For this reason, the preexponential factors are proportional to the ratio of the number of states available:

$$\frac{A_f}{A_r} \propto \frac{\Omega_{\text{propene}}}{\Omega_{\text{cyclopropane}}} \tag{4.46}$$

Combining Equations 4.45 and 4.46 gives

$$
\begin{aligned}
K &\approx \frac{\Omega_{\text{propene}}}{\Omega_{\text{cyclopropane}}} \frac{\exp\left(-(E_{\text{propene}} - E_{\text{cyclopropane}})\right)}{k_B T} \\
&= \frac{\Omega_{\text{propene}}}{\Omega_{\text{cyclopropane}}} \exp\left(-\frac{\Delta E}{k_B T}\right)
\end{aligned}
\tag{4.47}
$$

So the equilibrium constant is a simple function of the difference in energy and the difference in number of available states between the reactants and the products. We can also understand Equation 4.47 by forgetting about the intermediate state, and just applying the Boltzmann distribution directly to the reactants and products.

For an isomerization reaction such as this one, the change in volume $\Delta V \approx 0$. In a more general reaction done under constant-pressure conditions, we would have to add the work done on the surroundings ($P\Delta V$ discussed in Section 3.2) to the energy difference between the reactants and products, and we would replace ΔE with the enthalpy difference $\Delta H = \Delta H + P \Delta V$. Now take the natural log of both sides of Equation 4.47, and convert Ω into the entropy using Equation 4.29:

$$
\begin{aligned}
\ln K &= \ln \Omega_{\text{propene}} - \ln \Omega_{\text{cyclopropane}} - \frac{\Delta H}{k_B T} \\
k_B T \ln K &= k_B T \ln \Omega_{\text{propene}} - k_B T \ln \Omega_{\text{cyclopropane}} - \Delta H \\
&= T(S_{\text{propene}} - S_{\text{cyclopropane}}) - \Delta H \\
&= T\Delta S - \Delta H \\
\Delta H - T\Delta S &= -k_B T \ln K
\end{aligned}
$$

The quantity $\Delta H - T\Delta S$ is also called the Gibbs free energy difference ΔG. Since ΔG in Joules per molecule is a very small number for chemical reactions, we usually multiply by Avogadro's number on the left side (which gives ΔG in Joules per mole) and the right side (which gives $R = N_{\text{Avogadro}} k_B$). This gives the famous relationship between equilibrium constants and free energy differences:

$$\Delta G = -RT \ln K \tag{4.48}$$

► **PROBLEMS** ►

4-1.* Use the binomial distribution to calculate the exact probability of getting 95 or more heads out of 100 coin tosses.

4-2. Show that 95% of the time, n_H after 10,000 coin tosses will be between 4902 and 5098.

4-3.* Find the probability of getting 50.2% heads out of 1,000,000 tosses, and show that it is the same as the probability of getting more than 52% heads after 10,000 coin tosses.

4-4. Many books and scientific programs tabulate the *error function* which is defined as

$$\mathrm{erf}(y) = \frac{2}{\alpha\sqrt{\pi}} \int_0^{y\alpha} \exp\left(\frac{-x^2}{\alpha^2}\right) dx$$

The parameter α adjusts the width of the curve, but does not change the value of the error function. This definition makes $\mathrm{erf}(0) = 0$ *and* $\mathrm{erf}(\infty) = 1$, but in some ways it is not particularly convenient; usually we would rather find the area between limits expressed as multiples of the standard deviation σ. The normalized area between 0 and $z\sigma$ is

$$\frac{1}{\sigma\sqrt{2\pi}} \int_0^{z\sigma} \exp(-x^2/2\sigma^2)\, dx$$

Express this integral in terms of the error function.

4-5.* Calculating statistically valid error bars after a small number N of measurements (for example, six measurement of the concentration) is an important application of statistics. Sometimes we can estimate the error in each measurement based on our knowledge of the equipment used, but more often the sources of error are so numerous that the best estimate is based on the spread of the values we measure.

With a small number of measurements, calculating valid error bars requires a more complex analysis than the one given in Section 4.3. The mean is calculated the same way, but instead of calculating the root-mean-squared deviation σ, we calculate the **variance** s:

$$\mathrm{mean} = \overline{C} = \sum_{i=1}^{N} \frac{C_i}{N}$$

$$\mathrm{variance} = s = \left\{ \sum_{i=1}^{N} \frac{(C_i - \overline{C})^2}{N-1} \right\}^{1/2}$$

The variance is the best guess for the standard deviation of the distribution. If the errors are assumed to be random with a normal distribution, then 95% confidence limits for the mean are obtained by the following expression:

$$\text{Error bars: } \pm \frac{ts}{\sqrt{N}}$$

where the value of t is obtained from the table below.

Number of Measurements	t (95% confidence)
2	12.706
3	4.303
4	3.182
5	2.776
6	2.571
7	2.447
8	2.365
9	2.306
10	2.262
20	2.093
30	2.045
∞	1.960

(a) Show that, in the limit of a very large number of measurements, this approach reduces to the same method given in Section 4.3.

(b) Why is there no entry in the table for one measurement?

(c) The concentrations of six bottles of labeled 100 mH HCl are found to be 99.62 mM, 101.81 mM, 100.40 mM, 99.20 mM, 100.89 mM, and 100.65 mM. The average is greater than 100 mM. Can we conclude, with 95% confidence, that the real average concentration (if we had averaged the concentrations in a very large number of bottles) is greater than 100 mM?

4-6. Suppose you open a 22.4 liter box to the atmosphere at $0°C$. According to the ideal gas law, one mole of gas occupies 22.4 liters under those conditions, so there should be one mole of gas ($\approx 6.02 \times 10^{23}$ molecules) in the box. Based on the arguments in this chapter, if you tried this experiment many times, and counted the actual number of gas molecules you captured each time, how much would you expect your answer to fluctuate?

4-7.* Compare the concentration of oxygen molecules in the air in Denver, Colorado (approximate altitude 1500 m) to the concentration at sea level. You may assume the temperature remains constant at $25°C$.

4-8. When pollsters quote "error bars" or "likely errors," they are actually quoting "95% confidence limits." If the percentages in the polls are around 50%, this reduces to a "coin toss" problem just like the ones discussed in this chapter. Suppose 100 people are asked to compare French and California wines, and the preferences are exactly evenly divided. You could report that 50% prefer French wine. What would be the "error bars," given the assumptions above?

4-9.* Mary and Jane each have two children. At least one of Mary's children is a boy; Jane's first child is a boy. Show that Jane is more likely than Mary to have two boys.

4-10. A group of 100 students (50 men, 50 women) are divided up completely at random into pairs for a chemistry laboratory. Each pair is all-female, all-male, or mixed. What are the relative probabilities of these three outcomes?

4-11.* Isotopic substitution reactions, such as

$$^{14}N - ^{14}N + ^{15}N - ^{15}N \Leftrightarrow 2\,^{14}N - ^{15}N$$

involve only extremely small energy differences, so the equilibrium is determined almost completely by statistical differences. Assume you have an equal number of ^{14}N and ^{15}N atoms, and that they combine at random to form the three possible kinds of molecules above. Find the expected relative concentrations of the molecules, and find the expected equilibrium constant (hint: it is *not* one. If you don't see this, do Problem 4-10 first).

4-12. On the TV game show *Let's Make a Deal*, the contestant is shown three curtains. Behind two of the curtains are cheap gifts; behind the third curtain is the Grand Prize. She selects one curtain. The show's host (who of course knows which curtain hides the Grand Prize) then opens a different curtain to show a cheap gift. Finally, he gives the contestant the right to either stay with her original choice, or switch to the remaining unopened curtain. Statistically, what should she do? Why?

4-13.* Assume that $A = 100\pm5$ and $B = 15.0\pm1.8$. Find the error bars for the quantities $A + B$, AB, and A^3B^2.

4-14. Explain why the fractional error bars for A^2 are larger than the fractional error bars for AB, even if A and B each have the same fractional random error.

► Chapter 5

Introduction to Quantum Mechanics

Commonsense is nothing more than a deposit of prejudices laid down by the mind before you reach eighteen.

Albert Einstein (1879–1955)

5.1 PRELUDE

The latter half of the nineteenth century was a time of intellectual triumph in the physical sciences. Most of the material contained in the first year of modern college physics courses was completely understood by then. Newton's laws had been rephrased in different mathematical forms which simplified even complicated many-body problems such as planetary motion. In addition, the description of electric and magnetic fields by Maxwell's equations was an essentially complete success—so much so that these equations and their consequences are the central focus of some *graduate* physics courses even today.

The systemization of chemistry was also well underway, propelled in large part by Mendeleev's development of the Periodic Table in 1869. Two postulates and an enormous number of careful experimental measurements played a crucial role in this work:

1. Avogadro proposed in 1811 that equal volumes of different gases contained the same number of "elementary particles" (molecules). However, this led to a seemingly unreasonable conclusion. Since two volumes of hydrogen gas combine with one volume of oxygen gas to create two volumes of water, Avogadro's hy-

pothesis implies that the molecules of oxygen must split into two identical parts. But the only known forces were gravity, which was far too small to hold atoms together, and Coulombic attraction, which never generates an attraction between like particles!

As a result of this concern and others, Avogadro's hypothesis was rejected for nearly half a century. Ultimately, however, Avogadro's hypothesis made it possible to assign *relative* atomic weights to most of the lighter elements (determining the *actual* weight of an atom requires determination of Avogadro's number, which was not accurately known until Millikan determined the charge and mass of the electron in 1909).

2. Dulong and Petit proposed in 1819 that for most materials which melt far above room temperature,

$$\frac{dE}{dT} \approx 25 \text{ J} \cdot \text{mol}^{-1} \cdot \text{K}^{-1} \quad \left(\begin{array}{l} \text{substance confined to} \\ \text{a constant volume} \end{array} \right) \quad (5.1)$$

This quantity (the *constant-volume molar heat capacity* c_v) was derived for monatomic gases in the last chapter. The **rule of Dulong and Petit** was combined with stoichoimetric measurements to determine atomic weights for the heavier elements (Problem 5-1). Mendeleev started with a ranking of the elements by weight and then deduced the form of the periodic table, leaving gaps (and re-arranging iodine and tellurium into their proper order) on the basis of chemical properties. Shortly thereafter, the discovery of the elements gallium and germanium, which filled holes in the table and had properties he predicted on the basis of the other elements in those columns, emphatically verified the logic of this approach.

Dramatic progress had been made in other aspects of chemistry as well. The thermodynamic quantities which describe chemical reactions (such as energy and entropy) had been placed on a theoretical footing by statistical arguments. The kinetic theory of gases, which will be discussed in Chapter 7, was essentially complete. Many important reactions still used today to synthesize complicated molecules were first demonstrated in those years, and are still cited in modern scientific papers.

All in all, the accomplishments of nineteenth century chemistry and physics were impressive by any standard. At the beginning of that century, scientists (natural philosophers) often seemed not much different from Latin scholars or poets. By the end of the century, scientific research was an economic (and military) factor.

In fact, however, there were warning signs that something was badly wrong. We will focus here on six sets of observations which are relevant to chemical processes, and which together played a major role in the demise of what we now call classical physical theory.

1. Any heated object glows. The color of the glow changes as the temperature increases—from dull red (an electric stove) to yellow to "white hot". By the end of the nineteenth century the **spectrum** (distribution of wavelengths) of light emitted by a heated object had been accurately measured throughout the visible and much of the infrared region—it is very broad, and peaks at shorter wavelengths as the temperature increases. A variety of theories were introduced to explain the experimental results. In the end, the only theory which fit the experimental data required a nonsensical assumption, and the only theory which required no assumptions gave a physically impossible result.

2. Not all objects emit a broad distribution of wavelengths. A large voltage across two electrodes in a hydrogen-filled tube causes a discharge, but the gas glows at only a small number of different frequencies (three visible lines). Other gases produce different patterns of lines, both in the emission spectra of hot atoms and in the absorption spectra of cold atoms. In fact, helium was discovered in the sun (because of its absorptions) long before it was found on earth! The wavelengths of the visible absorptions and emissions (400–700 nm) are about 1000 times larger than atomic dimensions, so the restrictions to certain frequencies could not be equivalent to, say, the tones produced by a violin string. What could cause this peculiar property?

3. The rule of Dulong and Petit was not only obviously successful in placing elements in their proper places in the Periodic Table; it also had an extremely simple theoretical justification. A crystal could be modeled as a collection of atoms, held together by restoring forces analogous to springs. Inside the crystal each atom could move in three different directions (x, y, z), and each direction of motion would stretch one spring and compress another. Because of the Boltzmann distribution, the kinetic energy of each atom then had to be $3k_BT/2$ (just as with a gas); for each spring, the potential energy had to equal the kinetic energy (see Section 3.3). The net prediction was that the energy of a crystal of N atoms had to be:

$$E = K \text{ (kinetic energy)} + U \text{ (potential energy)}$$
$$= \frac{3Nk_BT}{2} + \frac{3Nk_BT}{2} = 3Nk_BT \text{ (any solid)} \tag{5.2}$$

so the heat capacity is

$$\frac{dE}{dT} = 3Nk_B = 24.9 \text{ J} \cdot \text{mol}^{-1} \cdot \text{K}^{-1} \tag{5.3}$$

in perfect agreement with Equation 5.1.

However, the rule of Dulong and Petit fails for some common substances at room temperature, and it fails for *all* substances at low temperatures—(dE/dT) approaches zero as the temperature goes to 0.

4. We noted in the last chapter that the Boltzmann distribution accurately predicts the heat capacity of monatomic gases such as helium. It also readily predicts the heat capacity of polytomic gases. As discussed in Chapter 3, atomic motions in polyatomic molecules include three translations, which are handled by the kinetic theory of gases in the same way as a single atom. Thus the three translation degrees of freedom (x, y,and z) each contribute $k_BT/2$ to the total kinetic energy per molecule, or $RT/2$ per mole. N-atom molecules also have 2 different rotations plus (3N-5) vibrations if the molecule is linear, or 3 rotations plus (3N-6) vibrations if it is nonlinear. The Boltzmann distribution can be used to show that each rotation also contributes $RT/2$ per mole to the total kinetic energy. In addition, just as with solids, we predict that the potential and kinetic energies are equal for the vibrations in a harmonic oscillator, so each vibration should contribute a total of RT to the energy or R to the heat capacity. So the Boltzmann distribution predicts that c_v is given by:

Linear molecule, N atoms:

$$c_v = 5R/2 \left(\begin{array}{c} 3 \text{ translations } + \\ 2 \text{ rotations} \end{array} \right) + (3N - 5)(R) \left(\begin{array}{c} \text{vibrational} \\ \text{energy} \end{array} \right)$$

$$= (3N - (5/2))R$$

Nonlinear molecule, N atoms:

$$c_v = 3R \left(\begin{array}{c} 3 \text{ translations } + \\ 3 \text{ rotations} \end{array} \right) + (3N - 6)(R) \left(\begin{array}{c} \text{vibrational} \\ \text{energy} \end{array} \right)$$

$$= (3N - 3)R \tag{5.4}$$

Equation 5.4 *disagrees with experiments*, which show that c_v is temperature dependent and substantially smaller than this predicted value at room temperature. The experimental value of c_v for virtually all diatomic gases at room temperature is about $5R/2$, not $7R/2$, and increases with temperature. For larger molecules the deviation from Equation 5.4 is even more substantial. This major disagreement was well known to Boltzmann —who committed suicide in 1906, when questions about the validity of his life's work seemed most serious.

5. The Periodic Table was completely empirical. There was no obvious reason why the ninth (fluorine), tenth (neon) and eleventh (sodium) elements should have such vastly different properties, or why this pattern should repeat for the 17th, 18th and 19th elements (and again for the 35th, 36th and 37th elements).

6. Even after Avogadro's hypothesis was universally accepted, and all chemists wrote oxygen as O_2, there was still no justification for why two oxygen atoms should come together as a molecule. There was also no obvious reason why H_2O should be more stable than HO or HO_2.

It was easy to dismiss these problems as specialized imperfections in a theoretical framework which was obviously enormously successful. In fact, however, they led to what is arguably the greatest revolution in modern science. In just a few decades all of the preceding theory of mechanics, electricity and magnetism lost its universality. The modern description of atoms and radiation is far richer but more complex. The theory which ultimately evolved, known as *quantum mechanics*, is counterintuitive and often inelegant. Commonsense will often not help you here; you will find many of the conclusions in this chapter to be positively crazy, and the philosophical consequences are quite profound. But the bottom line is simple—quantum mechanics works. It has survived almost a century of rigorous experimental tests. To be sure, just as we can contrast nineteenth-century chemistry and physics with twentieth-century ideas, someday a still further understanding will evolve; but this fuller theory will have to include the results of quantum mechanics as a special case, at least in the size and energy limits which include atoms and molecules.

5.2 BLACKBODY RADIATION—LIGHT AS PARTICLES

5.2.1 Properties of Blackbody Radiation

Light which hits a surface can be absorbed, transmitted, or reflected; most of the time some combination of these possibilities occurs. An idealized object which would absorb all of the radiation at all wavelengths is called a *blackbody*. A black, non-glossy sheet of paper, or carbon black from a sooty flame, is a good approximation to a blackbody for visible radiation.

Every substance at a finite temperature continually radiates energy, with a distribution (spectrum) which depends on the surface temperature. Figure 5.1 shows these spectra for two different temperatures: 5800K (the surface temperature of the sun) and 3000K (the surface temperature of the tungsten filament in a household *halogen lamp*, which is a special kind of incandescent light bulb).

Blackbody spectra have the following properties:

1. The total power radiated per cm^2 of surface area is proportional to the fourth power of the temperature:

$$P_{tot} = \sigma T^4 \qquad (5.5)$$

The modern value of σ, which is called the Stefan-Boltzmann constant, is $\sigma = 5.67 \times 10^{-8}$ W $\cdot$ m^{-2} $\cdot$ K^{-4}. For example, treating the filament in a 3000K halogen lamp as a blackbody is a fairly good approximation; if you take a 100-watt lamp, and turn down the voltage so that the power dissipated is only 50 watts, the temperature of the filament will fall by a factor of $\sqrt[4]{2}$ to about 2600K.

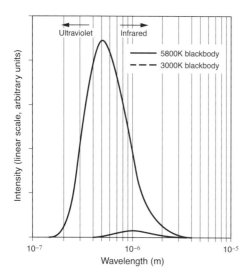

FIGURE 5.1 ▶ Energy density $U(\lambda)$ emitted as a function of wavelength for blackbodies at two different temperatures. The visible region ($\lambda \approx 0.4$–0.7μm) is shaded near the center. The total power radiated per unit area rises dramatically as the temperature increases. The spectrum shifts to shorter wavelengths as well.

2. The wavelength λ_{max} of maximum emission is inversely proportional to temperature:

$$\lambda_{\text{max}} T = 2898 \ \mu\text{m} \cdot \text{K} \tag{5.6}$$

For a 3000K halogen lamp $\lambda_{\text{max}} \approx 970$ nm, which is in the infrared region of the spectrum. Most of the radiated energy appears as heat, not visible light. The color of the light is also different from sunlight (it has relatively much more yellow and red), which is the reason that photographs taken with incandescent lighting seem to have distorted colors. The 2500K filament in a regular light bulb will be even less efficient in generating visible light, since λ_{max} will move farther from the visible.

Many different scientists tried to find the correct functional form for the experimental distribution $P(\lambda)$. By early 1900 the experimental data was good enough to rule out most of these attempts. In October 1900 Max Planck found a functional form which gave an excellent fit to experiment, but at that point the theoretical justification was quite weak. He worked for several years on a variety of derivations which relied on relations between thermodynamic properties. Ultimately, however, all of the derivations which gave the experimentally verified result were shown to require one assumption: the possible energies in each mode had to be restricted to discrete values $E = h\nu, 2h\nu, 3h\nu, \ldots$, where h was an arbitrary constant. In that case, Planck showed that the radiated energy density $U(\lambda)$ had to be:

$$U(\lambda) = \frac{8\pi hc}{\lambda^5(e^{hc/\lambda kT} - 1)} \tag{5.7}$$

This implied that for some reason the cavity radiated light only in packets (later called *photons*) with energy $h\nu$. A specific value of h (today called *Planck's constant*) gave perfect agreement with experiment. We also sometimes use frequency in units of radians per second (ω) instead of cycles per second (ν), as we discussed in Chapter 1, but we can convert readily between the two if we define $\hbar \equiv h/2\pi$:

$$E = h\nu = (h/2\pi)(2\pi\nu) = \hbar\omega \tag{5.8}$$

Both h and $\hbar$ (called "hbar") are widely used. The current best experimental values are $h = 6.6261 \times 10^{-34}$ J · s and $\hbar = 1.05457 \times 10^{-34}$ J · s.

It is important to understand how strange this looked. An analogy with sound waves is helpful: the power output (loudness) of a musical instrument certainly does not appear to be restricted to specific values, and flutes (high ν) do not appear to be restricted to a greater minimum loudness than tubas (low ν). Yet light waves did seem to have these properties. Naturally, such a strange result led others to work on the problem of blackbody radiation. Within a few years it became clear that the emitted spectrum could be directly calculated by a statistical treatment, without Planck's arbitrary assumption. The different frequencies and directions of emitted radiation are independent. Each of these independent "modes" of the radiation field can have any possible intensity. Of course very strong intensities in any mode are unlikely because of the Boltzmann distribution, but there are an infinite number of modes. The radiated energy density was predicted to be:

$$U(\lambda) = 8\pi kT\lambda^{-4} \tag{5.9}$$

Equation 5.9 is called the ***Rayleigh-Jeans law***, and can also be derived by taking the limit of Equation 5.7 as h approaches zero (Problem 5-2).

According to the Rayleigh-Jeans law, *all* objects gave off more blue light than red light, and more ultraviolet than visible light. Since the radiated power grows greater as the wavelength decreases, and wavelength can be arbitrarily short, this implied that everything radiates an infinite amount of power! The conflict between a completely reasonable derivation which gives an impossible result (the Rayleigh-Jeans law) and a derivation which requires a ridiculous assumption but gives the right answer in the laboratory (the Planck law) became generally understood in the physics community around 1908, and is called the *ultraviolet catastrophe*.

5.2.2 Applications of Blackbody Radiation

Planck's law is universally accepted today, and blackbody radiation is a tremendously important concept in physics, chemistry, and biology. The blackbody distribution is graphed on a log scale for a variety of temperatures in Figure 5.2.

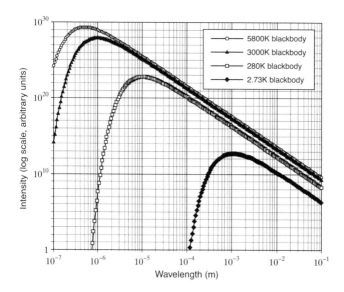

FIGURE 5.2 ▶ Blackbody radiation distribution for a variety of different temperatures. Notice that the curves shift with increasing temperature to shorter wavelengths and higher intensities, but otherwise they look identical.

We know that the surface temperature of the sun is approximately 5800K, because the spectrum of sunlight observed from outer space matches the distribution from a 5800K blackbody. At that temperature $\lambda_{max} \approx 500$ nm, which is blue-green light; perhaps coincidentally (but more likely not) the sensitivity of animal vision peaks at about the same wavelength. Unfortunately, this temperature is well above the melting point of any known material. The only practical way to sustain such temperatures is to generate sparks or electrical discharges. In fact one of the dangers of "arc welding" to join metals is the extremely high temperature of the arc, which shifts much of the radiated energy into the ultraviolet. The light can be intense enough to damage your eyes even if it does not appear particularly bright.

Tungsten filaments are the light source in incandescent light bulbs. The efficiency of such a bulb increases dramatically with increasing temperature, because of the shift in λ_{max}. Figure 5.3 illustrates this efficiency using a historical (but intuitive) unit of brightness—the candle.

The vapor pressure of tungsten also rises dramatically as the temperature increases, so increasing the temperature shortens the bulb life. In a standard light bulb, the operating temperature is held to about 2500K to make the lifetime reasonable (≈ 1000 hours). Halogen lamps, which have recently become widely available, incorporate a very elegant improvement. The filament is still tungsten, but a small amount of iodine is added. The chemical reaction

$$W(g) + 2I(g) \leftrightarrows WI_2 \tag{5.10}$$

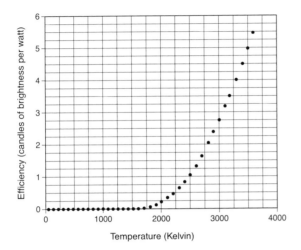

FIGURE 5.3 ► Efficiency of an ideal blackbody radiator for generating visible light (expressed as brightness per radiated watt). As the temperature increases, so does the fraction of light emitted in the visible. Thus the efficiency rises as well.

shifts back towards the reactants as the temperature increases. Near the walls of the quartz bulb the temperature is relatively cool, and tungsten atoms emitted by the filament react with the iodine to form WI_2 and other tungsten compounds. As these molecules migrate though the bulb they encounter the much hotter filament, which causes them to decompose—redepositing the tungsten on the filament and regenerating the iodine vapor. So halogen bulbs can run hotter (3000–3300K), yet still have a long life.

At still lower temperatures little of the emission is in the visible, but the effects of blackbody radiation can still be very important. The Sun's light warms the Earth to a mean temperature of approximately 290K; the Earth, in turn, radiates energy out into space. For the Earth $\lambda_{max} \approx 10 \ \mu$m, far out in the infrared. If this radiation is trapped (for example, by molecular absorptions) the Earth cannot radiate as efficiently and must warm. This is the origin of the *greenhouse effect*; as we will discuss in Chapter 8, carbon dioxide and other common gases can absorb at these wavelengths, so combustion products lead directly to global warming.

Even the near-vacuum of outer space is not at absolute zero. The widely accepted "Big Bang" theory held that the universe was created approximately 15 billion years ago, starting with all matter in a region smaller than the size of an atom. Remnants of energy from the initial "Big Bang" fill the space around us with blackbody radiation corresponding to a temperature of 2.73K. Detection of this "cosmic background" garnered the 1965 Nobel Prize in Physics for Penzias and Wilson. Measurements from the Cosmic Background Explorer satellite showed "warm" and "cool" spots from regions of space 15 billion light-years away. These temperature variations (less than 10^{-4} degrees!) reflect structures which were formed shortly after the "Big Bang," and which by now have long since evolved into groups of galaxies. Very recent measurements are forcing some modifications to the conventional framework. Not only is the universe still expanding; the expansion is accelerating!

5.3   HEAT CAPACITY AND THE PHOTOELECTRIC EFFECT

Two papers by Albert Einstein ultimately led to acceptance of the idea of quantization of energy for radiation, and were central to the development of the quantum theory (ironically, in later years Einstein became the most implacable critic of this same theory). The first of these papers, in 1905, concerned the **photoelectric effect**. Light ejected electrons from a metallic surface if the light had a greater frequency than some threshold frequency v_0 which depended on the particular metal. The kinetic energy K of the emitted electrons was proportional to the excess frequency, $v - v_0$ (Figure 5.4). Only the number of emitted electrons, not the kinetic energy, increased as the intensity increased.

Einstein rationalized these observations with an assumption closely related to Planck's—that the light existed only in bundles (quanta) of energy hv. Then hv_0 represented the energy needed to overcome the potential energy of binding of the electron to the metal, and the leftover energy ($hv - hv_0$) appeared as kinetic energy. The slope of the line in Figure 5.4 gave an independent measure of h, which agreed with Planck's value.

To the modern experimentalist, a device based on the photoelectric effect called a **photomultiplier tube** provides the most dramatic demonstration of the reality of photons (Figure 5.5).

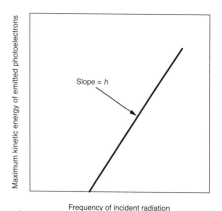

FIGURE 5.4 ▶ Light will only eject electrons from a metallic surface if the frequency is higher than a specific threshold, which varies for different metals. The maximum kinetic energy of the emitted electrons will increase if the frequency of the incident light increases. Einstein interpreted these observations as evidence for photons with energy $E = hv$.

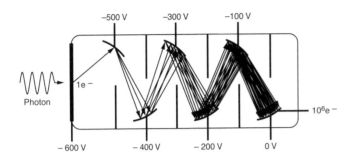

FIGURE 5.5 ▶ Schematic illustration of a photomultiplier tube. A single photon ejects an electron from the photocathode. The electron is accelerated by voltage differences, and knocks multiple electrons off each successive surface. The burst of electrons is collected at the end.

A photon kicks an electron off the first surface; the electron is accelerated by the voltage difference to the next electrode. The accelerated electron hits with sufficient force to knock off several more electrons, which in turn are accelerated to the next surface. After six or more of these stages, the cascade started by a single photon has become a burst of typically 10^6 electrons (total charge $Q \approx 10^6 e = 1.6 \times 10^{-13} C$).

This is still an extremely small charge; however, all of these electrons leave the photomultiplier tube within about 10 ns, because their velocities were determined almost completely by the accelerating voltages. Under *extremely* low light conditions, a measuring device such as oscilloscope will show a signal with intermittent bursts, each representing a cascade started by a single photon, on a baseline which has a small amount of noise. An electronic device called a "counter" simply examines this voltage continuously, and adds 1 every time the level exceeds a preset threshold. This type of **photon counting** system is used in hundreds of research laboratories to measure signals as small as a few photons per second (Problem 5-16), and with care it can be taken down to still much lower light levels.

Another paper of Einstein's showed that quantization of energy also predicted that Dulong and Petit's heat capacity rule would only be valid at high temperatures. Assume that the only allowed energies are $E = 0, h\nu, 2h\nu, \ldots nh\nu$, where n can be arbitrarily large. The average energy is

$$\langle E \rangle = \frac{(0)n_{E=0} + (h\nu)n_{E=h\nu} + (2h\nu)n_{E=2h\nu} + \cdots}{n_{E=0} + n_{E=h\nu} + n_{E=2h\nu} + \cdots} \tag{5.11}$$

(The denominator in this expression is the total population in all states; the numerator is the total energy). At very low temperatures ($k_B T \ll h\nu$) very little population is found in excited states, and the average energy is far less than $k_B T$. For example, consider the case $h\nu = 100 k_B T$. From the Boltzmann distribution, even the lowest excited state (with energy $E = 100 k_B T$) is almost empty:

$$n_{E=100k_BT} = n_{E=0} \exp\left(\frac{-(100k_BT)}{k_BT}\right) \approx 10^{-43} n_{E=0}$$

$$n_{E=200k_BT} \approx 10^{-86} n_{E=0}$$

If we save only the largest terms in the numerator and denominator of Equation 5.11 (or use the exact expression, given in Problem 5-5) we find:

$$\langle E \rangle \approx \frac{(100k_BT)(10^{-43}n_{E=0})}{n_{E=0}} = 10^{-41}k_BT$$

Thus vibrations with frequencies ν much higher than k_BT/h contribute essentially nothing to the total energy or to the heat capacity dE/dT. As the temperature approaches zero, *all* of the vibrational modes have frequencies much higher than k_BT/h, and thus they cease to contribute to the heat capacity.

At the opposite extreme, if $k_BT \gg h\nu_{max}$ (where ν_{max} is the highest vibrational frequency of the material) then there is a nearly continuous distribution of available energies. For example, the state with $E = k_BT$ has about 37% as much population as the state with $E = 0$; the state with $E = 2k_BT$ has about 14% of the population of $E = 0$. If you sum over all of the possible states, it can be shown that the classical $E = k_BT$ per vibration is recovered (Problem 5-5). Hence the rule of Dulong and Petit must be the high-temperature limit for all substances.

Einstein published this paper to explain measurements of the heat capacity of exactly one substance (diamond); almost nobody else in the scientific community thought there was anything wrong with the law of Dulong and Petit! But further measurements quickly verified that heat capacity did depend on temperature, and later refinements of this approach (most notably by Debye) gave excellent agreement with experiment. In a sense, chemists were just very lucky that "room temperature" is in the high temperature limit for most substances—or else the Periodic Table might have been deduced far later.

Heat capacities of polyatomic molecules can be explained by the same arguments. As discussed in Chapter 3, bond-stretching vibrational frequencies can be over 100 THz. At room temperature $k_BT \ll h\nu$ and these stretches do not contribute to the heat capacity (which explains why most diatomics give $c_v \approx 5R/2$, the heat capacity from translation and rotation alone). Polyatomic molecules typically have some very low-frequency vibrations, which do contribute to the heat capacity at room temperature, and some high-frequency vibrations which do not.

Extremely low temperature investigations (millikelvins or lower) remain an active area of modern physical research. The major motivation is often the decrease in the Boltzmann factor which makes it possible to put virtually all of the molecules in the lowest energy state, even if the energy ΔE to higher levels is very small. The simplest way to get stable low temperatures in the laboratory is to use commercially available condensed gases such as liquid nitrogen (which boils at 77K at atmospheric pressure) or liquid helium (which boils at 4K at atmospheric pressure, and about 2K at the lowest feasible pressures produced by a vacuum pump). These gases are prepared using refrigeration techniques such as those described in Chapter 7. Lower temperatures are usually achieved by putting additional equipment inside a cavity which is already cooled by liquid helium. But all of these techniques are dramatically complicated by the decrease in heat capacity as temperature approaches zero. A very tiny amount of energy coupled in by blackbody radiation from the room, or even blackbody radiation from the 4K liquid helium, is very effective in increasing temperature.

5.4   ORBITAL MOTION AND ANGULAR MOMENTUM

The motions of two bodies connected by an attractive force can also include rotation. In the simplest case (for example, rotation of the Earth around the Sun) one of the two bodies is much more massive than the other, and the heavier body hardly moves. The attractive force causes an acceleration through Newton's Second Law, but this does not necessarily imply that the speed changes—for a perfectly circular orbit the speed is constant. Velocity is a *vector* quantity, and so a change in direction is an acceleration as well.

Bound orbits due to the attraction between unlike charges (or due to gravity, which has the same functional form) are all circles or ellipses. We will consider the circular case first. Suppose a particle of mass m is moving counterclockwise in a circular orbit of radius R in the xy-plane. The particle's position $r = (x, y, z)$ can be described by the equations

$$x = R\cos(\omega t); \quad y = R\sin(\omega t); \quad z = 0$$
$$|\vec{r}| = \sqrt{x^2 + y^2 + z^2} = R \tag{5.12}$$

At $t = 0$ the particle is at the coordinate $(R, 0)$; at $\omega t = \pi/2$ the particle is at $(0, R)$; at $\omega t = \pi$ the particle is at $(-R, 0)$; at $\omega t = 3\pi/2$ the particle is at $(0, -R)$, and at $\omega t = 2\pi$ the particle has returned to its starting point.

The instantaneous velocity and acceleration are found by taking derivatives:

$$v_x = dx/dt = -\omega R \sin(\omega t)$$
$$v_y = \omega R \cos(\omega t)$$
$$v_z = 0$$
$$|\vec{v}| = \sqrt{v_x^2 + v_y^2 + v_z^2} = \omega R \tag{5.13}$$
$$a_x = d^2x/dt^2 = -\omega^2 R \cos(\omega t)$$
$$a_y = -\omega^2 R \sin(\omega t)$$
$$a_z = 0$$
$$|\vec{a}| = \sqrt{a_x^2 + a_y^2 + a_z^2} = \omega^2 R \tag{5.14}$$

Notice that the acceleration vector is pointed directly toward the origin:

$$\vec{a} = -\omega^2 \vec{r}; \quad |\vec{a}| = \omega^2 R \tag{5.15}$$

According to Newton's laws, this acceleration can only be sustained if there is an attractive force $\vec{F} = m\vec{a}$ towards the origin:

$$\left|\vec{F}\right| = m\,|\vec{a}| = m\omega^2 R = -\frac{q_1 q_2}{4\pi\varepsilon_0 R^2} \quad \text{(Coulomb's law, circular orbit)} \tag{5.16}$$

The minus sign comes into Equation 5.16 because $q_1 q_2 < 0$ (the charges have opposite signs) for a bound orbit. If the initial distance R from the origin is specified, Equation 5.16 implies that the rate of rotation ω is also fixed, as is the speed ωR. Rearranging Equation 5.16 gives:

$$\omega = \sqrt{-\frac{q_1 q_2}{4\pi\varepsilon_0 R^3 m}} \tag{5.17}$$

Starting from this result we can readily find the speed and energy:

$$\text{Speed } s = |\vec{v}| = \omega R = \sqrt{-\frac{q_1 q_2}{4\pi\varepsilon_0 R m}}$$

$$\text{kinetic energy } K = \frac{m s^2}{2} = -\frac{q_1 q_2}{8\pi\varepsilon_0 R}$$

$$\text{potential energy } U = \frac{q_1 q_2}{4\pi\varepsilon_0 R}$$

$$\text{total energy } E = K + U = \frac{q_1 q_2}{8\pi\varepsilon_0 R} \tag{5.18}$$

Note that these equations only make any sense if $q_1 q_2 < 0$ (the charges are of opposite sign, giving an attraction), so the kinetic energy is positive as expected.

Circular orbits are a special case. More generally orbits are ellipses, and then the speed is not constant. The velocity can be decomposed into components parallel

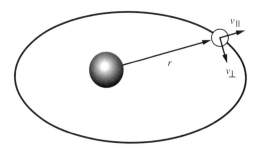

FIGURE 5.6 ▶ For elliptical orbits, the sum of kinetic and potential energy is conserved, as is the quantity $\left|\vec{L}\right| = m v_\perp r$ (angular momentum). The orbiting body moves fastest at the minimum separation.

and perpendicular to the position vector (Figure 5.6). The component parallel to r, called $v_\parallel$, changes the length of the position vector (if $v_\parallel > 0$ the separation is increasing) so for a circular orbit $v_\parallel = 0$. The other component, perpendicular to r, is called $v_\perp$. For an elliptical orbit, $v_\parallel = 0$ only when $|\vec{r}|$ is at a minimum or maximum. Comets have highly elliptical orbits; planetary orbits are more nearly circular. For example, the Earth's distance to the sun varies by 3.2% over a year, as does its speed.

Newton's laws can also be used to describe elliptical orbits, and it is then found that a vector quantity called *angular momentum* is conserved (always stays constant), just as the total momentum $\vec{p}$ stayed constant in a closed system. Angular momentum is conventionally denoted $\vec{L}$. The vector $\vec{L}$ points perpendicular to the orbit (in the z-direction by our definitions) and has length

$$\left|\vec{L}\right| = mv_\perp r = \text{constant} \tag{5.19}$$

For a circular orbit with radius R, since $v_\parallel = 0$ and $v_\perp = \omega R$, Equation 5.17 gives

$$\left|\vec{L}\right| = m\omega R^2 = \sqrt{-\frac{q_1 q_2 R m}{4\pi \varepsilon_0}} \tag{5.20}$$

$$\text{kinetic energy } K = \frac{mv^2}{2} = \frac{m\omega^2 R^2}{2} = \frac{\left|\vec{L}\right|^2}{2(mR^2)} \equiv \frac{\left|\vec{L}\right|^2}{2I} \tag{5.21}$$

The quantity $I = mR^2$ in the denominator of Equation 5.21 is called the *moment of inertia*, and is labeled I.

Just as in the case of vibrational motion, the equations become a little more complex if the two objects have similar masses. It can be shown that Equations 5.20 and 5.21 above still hold, if m is replaced by the reduced mass $\mu = m_1 m_2/(m_1 + m_2)$:

$$\left|\vec{L}\right| = \mu\omega R^2, \quad \mu = m_1 m_2/(m_1 + m_2)$$

$$K = \frac{\left|\vec{L}\right|^2}{2I}, \quad I = \mu R^2 \text{ (circular orbit, two connected masses)} \tag{5.22}$$

The concept of angular momentum was introduced centuries ago, and might seem rather specialized. In fact, however, **it is impossible to overstate the importance of angular momentum in chemistry**. We will show in later sections that electrons orbit the nuclei of atoms or molecules with trajectories which are far more subtle than circular or elliptical orbits, but the concept of angular momentum still applies. It is central to bonding; it explains why element 10 (neon) is stable while elements 9 (fluorine) and 11 (sodium) are highly reactive; and it lets us predict what combinations of atoms will form stable molecules. Nuclei and electrons themselves have angular momentum, as if they were spinning, and the properties of these "spins" are probed countless times a day in chemistry departments to measure molecular structures. Finally, the molecule as a whole has angular momentum, and this property is used to measure bond lengths and angles (and to heat food in a microwave oven).

5.5 ATOMIC STRUCTURE AND SPECTRA-QUANTIZATION OF ENERGY

If we accept that light exists as photons, then the presence of specific and sharp frequencies in the emission spectra of atoms must be interpreted as restricting the internal energy of atoms to specific values. For single-electron atoms the frequencies of the observed photons satisfy the simple relationship

$$v = Z^2 (3.28984 \times 10^{15} \text{ Hertz}) \left(\frac{1}{m^2} - \frac{1}{n^2} \right) \tag{5.23}$$

where m and n are arbitrary integers and Z is the atomic number ($Z = 1$ for hydrogen, $Z = 2$ for He$^+$, and so forth). The results of Section 5.3 imply that we should assign an energy $E = hv$ to each photon emitted at frequency v. Conservation of energy would then imply that production of a photon requires a change in the internal energy of the atom. A transition from a state with energy E_m to a state with energy E_n then produces a photon with energy $hv = E_m - E_n$.

$$E = hv = Z^2 (2.1799 \times 10^{-18} \text{ Joules}) \left(\frac{1}{m^2} - \frac{1}{n^2} \right) = E_m - E_n \tag{5.24}$$

When m gets very large, $1/m^2$ approaches zero. So if we pick the zero of energy to correspond to the limit of very large m, we find:

$$E_n = (-2.1799 \times 10^{-18} \text{ Joules}) Z^2 / n^2 \tag{5.25}$$

The energy required to take a hydrogen atom from $n = 1$ (the lowest state) to $n = \infty$ is 2.1799×10^{-18} Joules, which is called the *ionization energy*.

The first real clues about the structure of atoms came in 1909 when Rutherford bombarded a thin film of gold with α-particles (helium nuclei). The deflection pattern was consistent only with an atomic structure which had the positive charge and most of the mass concentrated at the center. This immediately suggested a planetary model (called the **Bohr model** after Niels Bohr, who received the 1922 Nobel Prize in Physics for this work), with electrons orbiting the nucleus, held in place by the attraction between unlike charges. The expected properties of an electron (mass m_e charge $-e$) orbiting a nucleus (mass m_n charge $+Ze$) at radius R were derived in the last section. For a circular orbit there are three conserved quantities, or constants of the motion: the total energy (Equation 5.18), the length of the angular momentum vector (Equation 5.20), and the direction of the angular momentum vector (perpendicular to the orbit). Equating the energy in Equation 5.18 to the observed result in Equation 5.25 restricts the possible values for R:

$$E = \frac{-Ze^2}{8\pi \mathcal{E}_0 R} = \frac{Z^2 (-2.1799 \times 10^{-18} \text{ Joules})}{n^2} \tag{5.26}$$

The different values of n in Equation 5.25 then correspond to orbits with different radii, angular momenta, and rotation rates. For example, for hydrogen ($Z = 1$), we would get

$$R(\text{Bohr model}) = \frac{n^2 e^2}{8\pi \varepsilon_0 (2.1799 \times 10^{-18} \text{ Joules})} = n^2 (52.92 \text{ pm})$$

$$(5.27)$$

Substituting Equation 5.27 into the expression for the angular momentum (Equation 5.20) gives:

$$\left| \vec{L} \right| (\text{Bohr model}) = n(1.0546 \times 10^{-34} \text{ J} \cdot \text{s}) \qquad (5.28)$$

The numerical constant in Equation 5.28 is the same as $\hbar$ in the Planck relationship. So the assumption that *electrons have circular orbits with angular momentum restricted to multiples of $\hbar$* gave the correct emission spectra for hydrogen. We could also start from the assumption $\left| \vec{L} \right| = n\hbar$, rearrange Equation 5.20, and derive R:

$$R = \frac{4\pi \varepsilon_0 \left| \vec{L} \right|^2}{me^2} = n^2 \frac{4\pi \varepsilon_0 \hbar^2}{m_e e^2} \qquad (5.29)$$

In honor of Bohr, the value of R for $n = 1$ is called the *Bohr radius* a_0:

$$a_0 = \frac{4\pi \varepsilon_0 \hbar^2}{m_e e^2} = 52.92 \text{ pm} \qquad (5.30)$$

In one very important way, circular orbits held together by Coulomb's law are different from orbits held together by gravity. The electron moves at constant speed, but its direction is changing, therefore it is accelerating (this is also a direct consequence of $\vec{F} = m\vec{a}$). However, an accelerating electron radiates electromagnetic energy. So energy conservation implies that the total energy must constantly decrease, and the electron must spiral down very quickly to the nucleus. So once again we have an extremely strange result which agrees perfectly with experiment. If we take it at face value, electrons can only stay in orbits which give them specific values of angular momentum, and if they are in these orbits they do not radiate energy. But these create still more questions. For example, why should the only orbits be circular? Wouldn't a hydrogen atom look different from a direction perpendicular to the orbit than it does from other directions? Finally, many attempts were made to extend these ideas to larger atoms or to molecules, and they all failed miserably.

The Bohr model was an important step forward in understanding atomic structure, but just because it was recognized by a Nobel Prize *does not mean it is correct!* As we show in the next few chapters, the hydrogen atom energy is indeed restricted to the values given by Equation 5.25, and the Bohr radius a_0 will reappear as a convenient parameter in the correct solution for the hydrogen atom. However, Equation 5.28 does not give the correct restriction on the length of the angular momentum vector, and electrons do not "orbit" the nucleus.

5.6 PARTICLES AS WAVES

Yet another step towards the new theory was taken in 1924 by Louis de Broglie, a graduate student at the time at the Sorbonne in Paris. So far we have seen that energies of matter (hydrogen atoms) and light are both quantized, and that light can be shown to behave sometimes as particles. Can particles behave sometimes as waves?

Wavelike behavior, such as *interference*, depend directly on the wavelength, and it is far from obvious what "wavelength" you would assign to a hydrogen atom, let alone a macroscopic object such as a baseball. For photons we have

$$E = h\nu = hc/\lambda \text{ (photons)} \tag{5.31}$$

In addition, classical electromagnetic theory predicts that light waves have momentum given by

$$p = E/c \tag{5.32}$$

so we can combine these two equations to write

$$p = \frac{h}{\lambda} \text{ (photons)} \tag{5.33}$$

de Broglie argued that Equation 5.33 could be used to define the *wavelength* of particles:

$$p = \frac{h}{\lambda} \text{ (particles)} \tag{5.34}$$

Of course, the definition might not be very useful; we could also "define" the ears of an elephant to be wings, but this would not make elephants fly. What makes this definition interesting is a very simple extension. He also argued that the orbit of an electron around a proton would only be stable if its circumference $2\pi R$ were an integral number of wavelengths. Thus

$$n\lambda = \frac{nh}{p} \text{ (from Equation 5.34)} = 2\pi R \tag{5.35}$$

Rearranging this gives $pR = n(h/2\pi) = n\hbar$, but $\left|\vec{L}\right| = mv_\perp R = pR$ for a circular orbit (see Equation 5.19), so this implies $\left|\vec{L}\right| = n\hbar$. Thus Bohr's restriction of angular momentum to multiples of $\hbar$ is exactly the same as de Broglie's assumption that electrons have a wavelength which determines the allowed orbits.

5.7 THE CONSEQUENCES OF WAVE-PARTICLE DUALITY

de Broglie's successful description of the hydrogen atom with his hypothesis that particles could have wavelike behavior is not just a mathematical curiosity; it has tremendous experimental consequences. The Bohr model is obsolete, but the concept of a *de*

Broglie wavelength (Equation 5.34) is not. Countless experiments have shown that particles really *do* have a wavelength.

The most straightforward way to understand the consequences of the wave-particle duality is to return to an application we first discussed in Chapter 3—examining the effects of combining two identical sources. As Figure 3.6 showed, the distribution from two sources of classical particles (for example, two shotguns) is merely the sum of the distributions from each source; the distribution from two sources of classical waves (for example, two speakers playing the same tone) has superimposed interference patterns. For particles, the *intensity* of each source (which is never negative) adds to produce the net intensity; for waves, the *amplitude* of the wave (which can be positive or negative) adds, so destructive or constructive interference is possible.

What we have shown, however, is that light and matter can have both wavelike and particle-like character. So let us generalize Figure 3.6 to include two other sources: a light source (Figure 5.7) and a beam of electrons (Figure 5.8). An intense monochro-

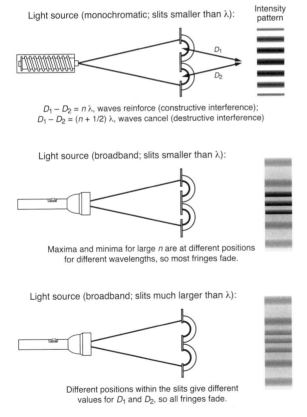

FIGURE 5.7 ► Comparison of interference fringes from two wave sources under different conditions. Notice that the fringes fade if the light is not monochromatic, or if the slits are large.

matic light source, such as a laser, produces an interference pattern from two small slits which is essentially the same as the sound pattern in Figure 3.7; in fact, as we noted in Chapter 3, the wavelike character of X-rays (electromagnetic fields with wavelengths comparable to crystal lattice spacings) permits measurement of distances in crystals because of the interference pattern.

Interference is an important aspect of wavelike behavior, but it is not always obvious. For example, two flashlights do not produce an interference pattern on a wall; their intensities appear to add. The disappearance of fringes comes from several effects. The two sources are large compared to an optical wavelength; the output of a flashlight has a broad range of colors; and the waves from two flashlights do not have a well-defined phase relation between them. However, even a complex source such as a flashlight will generate interference under the right conditions. Young first explored interference in 1801, and Michaelson and Morley used interference between light waves from one source traveling along two nearly equal paths to attempt to measure changes in the speed of light in 1887.

If the intensity is turned down dramatically, so that only a small number of photons go through the slits at a time, it is possible to resolve individual photons. A photomultiplier tube positioned near the first maximum would eventually detect a large number of photons; a tube positioned near the first minimum would detect fewer photons.

Now let us examine a beam of small particles such as electrons. Since $\lambda = h/p$, we know that the prescription for maximizing interference will be to prepare this beam with all of the electrons having nearly the same momentum. This can be done with a ***cathode ray tube*** (CRT), which is simply an evacuated tube with a very high voltage difference between two electrodes (Figure 5.8). Electrons are attracted to the more positively charged plate (the anode). If the voltage difference is high enough, it can overcome the binding energy of the electrons to the metal in the cathode, and electrons travel from the cathode to the anode. These electrons are accelerated by the potential difference.

Suppose there are two extremely small holes in the anode. The electrons passing through these holes can strike the end of the tube, which is coated with a *phosphor*—a

Electron beam (particles, wavelength comparable to slit spacing):

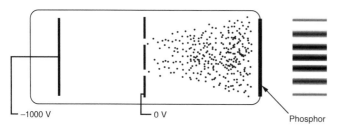

FIGURE 5.8 ▶ Simplified illustration of a cathode ray tube. The "cathode rays" are electrons which can be pulled off the cathode by a large potential difference between cathode and anode.

material which glows when electrons hit it. The spatial distribution of the glow will show *interference effects from the two different electron paths*. The positions of the maxima and minima can be predicted (just as for light waves) by determining whether the path length difference is an integral or half-integral multiple of wavelengths.

All of the other phenomena associated with waves can also be observed in particles. For example, in 1927 Davisson and Germer accelerated a beam of electrons to a known kinetic energy and showed that these electrons could be diffracted off a nickel crystal, just as X-rays are diffracted (see Figure 3.8). Just as with photons, interference is not always seen: if the wavelength spread or the slits are large, the fringes wash out. This also explains why interference is not seen with macroscopic objects, such as buckshot—the wavelength is far too small.

In general, the tradeoff between particle-like and wave-like properties depends on the *spatial resolution of the measurement*. If this resolution is much larger than the wavelength, interference effects disappear. For example, in the photoelectric effect light ejects an electron from a surface whose area is much larger than one square wavelength; the surface is also generally smooth on the scale of one wavelength ($\approx 1\mu$m). As a result, the behavior is dominated by particle-like properties, as discussed earlier. Interferometric measurements require observation of fringes from path length differences comparable to a wavelength. Similarly, *electron diffraction* comes from beams of energetic electrons with a wavelength comparable to the lattice spacing. On the other hand, television sets use cathode ray tubes with a beam diameter much greater than the electron wavelength, and interference effects wash out.

5.8 CLASSICAL DETERMINISM AND QUANTUM INDETERMINACY

Quantum mechanics has tremendous philosophical consequences which are still debated to this day, and which go well beyond the scope of this book. Perhaps the most important of these consequences is the destruction of *classical determinism*, and the recognition that it is impossible to make observations without fundamentally changing the system being observed. This result is quantified by the Heisenberg Uncertainty Principle, which is simply a consequence of the wavelike nature of matter.

5.8.1 Classical Uncertainty: Predicting the Future

Newton's laws are perfectly *deterministic*. Suppose we have a set of masses, all of whose positions and momenta are specified at some instant in time. Further, suppose that we specify all of the forces of interaction between these balls. We could now use Newton's laws to determine, *exactly*, the state of this system at any later time.

For example, suppose we knew that at time $t = 0$, a ball was centered in a box with length 0.5 meter, and suppose we knew that it was moving to the right at exactly $1 \text{ m} \cdot \text{s}^{-1}$. In the absence of any forces, we could predict exactly where it would be next week, next month, or next year. The round-trip distance is one meter, so the ball returns to its original position (moving to the right) at exactly $t = 1s$, $t = 2s$, and so forth.

Of course, there are uncertainties in any realistic measurement, and this ultimately limits our ability to predict the future position of the ball. Suppose we only knew the initial velocity to 1% accuracy: $v = 1.000 \pm 0.01$ m · s^{-1}. Then the uncertainty in the total distance traveled ($x = vt$) grows with time (Figure 5.9):

$$t = 10 \text{ sec}, x = 10.00 \pm 0.1 \text{ m } (10.00 \pm 0.1 \text{ round trips});$$
$$t = 100 \text{ sec}, x = 100.0 \pm 1 \text{ m } (100.0 \pm 1 \text{ round trips})$$

For short times we know the approximate position, but at long times the position is essentially unknown. However, there is no fundamental physical limit to our ability to measure the velocity; so, in principle, we can do the initial measurement with enough accuracy to predict the position at any later time. For example, we could accurately predict the position of the ball after one year (31,557,600 seconds) if we measured the initial velocity with an uncertainty much better than one meter in 31,557,600 seconds.

According to Newtonian physics, the universe and all of the objects in it are simply an extremely complicated collection of masses and charges. This implies that if you knew the state of the universe at any one time (exactly), you could predict (exactly) the future. Of course it is unrealistic to actually *do* a set of measurements which define the conditions at one time with enough accuracy, but it is *in principle* possible. *Therefore, Newtonian mechanics predicts that the future is perfectly determined by the past.* There is no such thing as "free will", and you need not worry about studying for the next examination; it has already been determined how much you will study, and what you will score on the test!

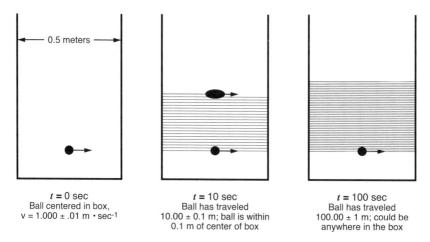

$t = 0$ sec	$t = 10$ sec	$t = 100$ sec
Ball centered in box,	Ball has traveled	Ball has traveled
v = 1.000 ± .01 m · sec⁻¹	10.00 ± 0.1 m; ball is within	100.00 ± 1 m; could be
	0.1 m of center of box	anywhere in the box

FIGURE 5.9 ▶ Uncertainty in position, coming from uncertainty in initial knowledge of velocity. This uncertainty can be reduced to an arbitrarily small value by a sufficiently accurate initial measurement.

5.8.2 The Crushing Blow to Determinism

Up to now, we have summarized quite a number of seemingly bizarre results. However, all of these results can be understood by accepting the idea of wave-particle duality, which may not be predicted from classical physics, but is not intrinsically inconsistent with it. It *could* be argued that now we simply know more about the nature of matter and light; thus the concepts of waves and particles, previously thought to be so different, simply have to be extended to a middle ground.

There is still a problem with such a compromise. To have interference, waves must be simultaneously present from two sources, so that they can cancel or reinforce. Yet if the intensity of the light in Figure 5.7 is turned down very low (so that only one photon is present at a time), or if the intensity of the electron beam in Figure 5.8 is decreased (so that only one electron hits the phosphor at a time), *fringes still build up with time.* The electron or photon arrives at different positions with probabilities that exactly mirror the fringes observed at high intensity.

How can this be? If the electron passed through the top slit in Figure 5.10, and no other electrons are present in the tube at the time, how can it *matter* that a second slit also exists? Similarly, if it passed through the bottom slit, why would the top slit affect anything? Therefore, how can we get interference?

An experimentalist, puzzled by this unreasonable result, would quickly pose a modification to Figure 5.10 to clarify the issue: detecting which slit the single electron actually used. For example, a small coil could surround the upper slit. The moving electron would produce a current, and this would induce a voltage in the coil. Now detecting this voltage would determine if the electron used the upper or lower slit.

If we do this, the interference vanishes (Figure 5.11).

You might chalk this disappearance up to a badly designed experiment. There are many other ways to measure the position of an electron (for example, you can scatter light off it). But *every experiment ever devised destroys the interference as soon as it becomes possible in principle to distinguish between the slits.* You cannot know more

Low intensity electron beam (only one electron present at a time):

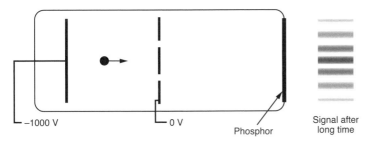

−1000 V   0 V
                   Phosphor   Signal after
                                       long time

FIGURE 5.10 ▶ If only one electron is present in the system at a time, but there are two possible and indistinguishable paths, the ultimate signal will show interference effects.

about the electron's trajectory without fundamentally changing the way it behaves! If it is not possible to distinguish the two trajectories, all of the properties of the system are the same as if *the particle went through both slits at the same time*.

5.8.3 The Heisenberg Uncertainty Principle

Figure 5.11 shows one of the most counterintuitive results of quantum mechanics. There are fundamental limits to our ability to make certain measurements—the act of determining the state of a system intrinsically perturbs it. For example, it is impossible to measure position and momentum simultaneously to arbitrarily high accuracy; any attempt to measure position automatically introduces uncertainty into the momentum. Similarly, a molecule which is excited for a finite period of time cannot have a perfectly well-defined energy. As a result, classical determinism fails. It is not possible, even in principle, to completely specify the state of the universe at any instant, hence the future need not be completely defined by the past. These results are usually phrased something like:

$$\Delta x \Delta p \gtrsim \frac{h}{4}; \quad \Delta E \Delta t \gtrsim \frac{h}{4} \tag{5.36}$$

and are examples of what is called the ***Heisenberg Uncertainty Principle***.

The Heisenberg Uncertainty Principle and its far-reaching consequences actually follow directly from the conclusion that particles also have wavelike character (and vice versa). Consider, for example, two waves with slightly different frequencies (Figure 5.12). The waves can be made to constructively interfere at one point, but eventually the difference in frequency will cause them to destructively interfere. In Figure 5.12 the waves are chosen to have a 10% frequency difference. So when the slower wave goes through 5 full cycles (and is positive again), the faster wave goes through 5.5 cycles (and is negative).

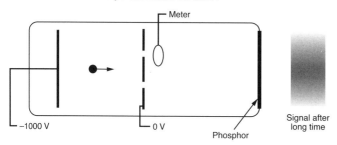

FIGURE 5.11 ▶ If it is possible to detect which of two interfering paths an electron actually takes, the interference vanishes—no matter how carefully the apparatus is designed to minimize perturbations.

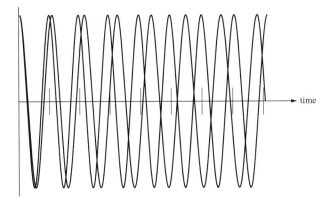

FIGURE 5.12 ▶ Two waves at different frequencies will constructively interfere and destructively interfere at different times.

Suppose that we had a very large number of sine waves, all arranged so that they were in phase at time $t = 0$, with a random (Gaussian) distribution of frequencies. Eventually the waves will start to destructively interfere (cancel), and the amount of time the waves remain in phase depends on the width of the frequency distribution. For a distribution of frequencies with an uncertainty in frequency of ±5% ($\Delta \nu = 0.05\nu_0$, where ν_0 is the average frequency), it can be shown that the sum looks like Figure 5.13.

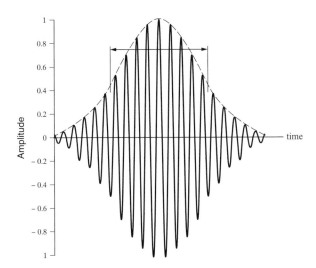

FIGURE 5.13 ▶ The sum of a large number of sine waves, with a distribution of frequencies of 5% around the center frequency, produces constructive interference for a range $\pm \Delta t \approx \pm 5$ cycles from the time when they all are in phase (the arrow in the figure). The length of the arrow is thus $2\Delta t$.

The waves produce a *pulse* of radiation with a finite width (represented by the arrow in the figure). We will define the uncertainty Δt as the half width at half maximum; thus we would use half the length of this arrow, or approximately five cycles at the center frequency ν_0. The pulse is large as long as most of the frequency components constructively interfere, then grows small. From the figure, Δt is approximately five cycles at the center frequency ν_0. Each cycle lasts for a time $1/\nu_0$. The net result is:

$$\Delta t \approx 5 \cdot (1/\nu_0); \quad \Delta \nu = 0.05\nu_0$$
$$\Delta \nu \Delta t \approx 1/4 \quad \text{(or equivalently } \Delta \omega \Delta t \approx \pi/2) \tag{5.37}$$

We used "$\approx$" instead of "=" in Equation 5.37 because the exact numerical value depends on the definition of the uncertainties—you will see different values in different books. If we define "Δt" in Figure 5.13 as the full width at half maximum or the root-mean-squared deviation from the mean, the numerical value in Equation 5.37 changes. It also changes a little if the distribution of frequencies is not Gaussian. Equation 5.37 represents the best possible case; more generally we write

$$\Delta \nu \Delta t \gtrsim 1/4 \tag{5.38}$$

Equation 5.38 turns out to be a universal result. If we increase the range of frequencies, then the waves constructively interfere for a shorter time. In order to have a single, well-defined frequency ($\Delta \nu \to 0$) the wave needs to continue for a very long time ($\Delta t \to \infty$).

If a wave persists only for a time Δt (or if we can only measure its frequency for a finite time Δt), *the frequency is intrinsically unknown by an amount* $(1/4\Delta t)$. For example, suppose we try to measure the frequency of a sound wave by using a microphone and an oscilloscope to count the number of cycles in one second. We could readily distinguish between a sound wave at 1000 Hz and one at 1001 Hertz, because the faster wave will go through one more cycle (in fact, your ear would hear a "beat" as the notes went in and out of phase with each other). It would be nearly impossible to distinguish between a wave at 1000 Hz and a wave at 1000.001 Hz; at the end of one second, they would still be nearly perfectly in phase with each other.

Planck's relationship $E = h\nu$ lets us convert Equation 5.38 into a relation between energy and measurement time.

$$\Delta E \Delta t \gtrsim \frac{h}{4} \tag{5.39}$$

Also, the equation which describes a sine wave moving in the x direction is

$$\mathcal{E}(x, t) = \mathcal{E}_0 \cos\left(2\pi \left(\nu t + \frac{x}{\lambda}\right)\right) \tag{5.40}$$

so the product $x \cdot (1/\lambda)$ must behave the same way as the product $\nu \cdot t$. Thus

$$\Delta x \Delta \left(\frac{1}{\lambda}\right) \gtrsim \frac{1}{4} \tag{5.41}$$

and since $p = h/\lambda$ (the de Broglie formula),

$$\Delta x \, \Delta p \gtrsim \frac{h}{4} \tag{5.42}$$

It is impossible to simultaneously specify the position and the momentum in one dimension to arbitrarily high accuracy. Any measurement which locates the position of a wave in time (or in space) guarantees that there must be a distribution of energies (or momenta). In the two-slit experiment, measuring the position of the electron (thus reducing the uncertainty Δx in position) introduces an uncertainty in the momentum and wavelength, and if Δx is small enough to determine which slit was used, the introduced uncertainty is always sufficient to eliminate the fringes.

5.9 APPLICATIONS OF THE UNCERTAINTY PRINCIPLE

The Uncertainty Principle often permits simple explanations of complex quantum mechanical results. Here are a few examples.

1. *Confining a particle to a restricted region in space increases its minimum possible energy.* Let us consider one particularly simple case—a particle in a one dimensional "box". The box is defined by a potential

$$U(x) = \left\{ \begin{array}{ll} 0, & 0 < x < L; \\ \infty, & x \leq 0 \text{ or } x \geq L \end{array} \right\} \tag{5.43}$$

This potential is the same as a flat-bottomed container with infinitely high walls separating inside from outside. Here we will use the Uncertainty Principle to estimate the minimum energy; later (in Chapter 6) we will find all of the possible energies and states for this system, using a differential equation known as *Schrödinger's equation*.

- The particle cannot sit motionless in the box. Such a state would have $p = 0$, hence $\Delta p = 0$ (there is no uncertainty). Since $\Delta x \, \Delta p \geq h/4$, this is only possible if Δx is infinite. But the particle is certainly in the box, so this is impossible.

- The energy of the particle is entirely kinetic energy, since the potential energy is zero inside the box. So a state with fixed energy E has a fixed value of $|\vec{p}| = \sqrt{2mE}$. The uncertainty comes from the fact that momentum is a *vector* quantity, and the momentum can be either positive or negative (the particle can be moving either right or left). So the uncertainty Δp is:

$$\Delta p = \sqrt{2mE} \tag{5.44}$$

- The maximum possible range of x is $\pm L/2$ about the middle. If we assume $\Delta x \approx L/2$ we predict

$$\Delta p \geq h(4\Delta x) = h/2L \qquad (5.45)$$

Combining Equations 5.44 and 5.45 gives

$$E \geq h^2/(8mL^2) \quad \text{(assuming } \Delta x \approx L/2) \qquad (5.46)$$

The lowest energy state decreases in energy as the box expands (enlarging the box permits a larger value of Δx, hence a smaller value of Δp). Thus, for example, a ball in a box must always be moving, but it can have a lower minimum speed if the box is big.

We could get the same answer in a different way, using de Broglie's relation $\lambda = h/p$ (Problem 5-15). The wave representing the electron would have to vanish at the two walls, similar to the waves on a violin string. The longest possible wave we could fit into the box would have wavelength $\lambda = 2L$. Such a wave would go through half a cycle between the two walls, and would be zero at each wall.

The minimum energy is trivially small for macroscopic objects, so these effects are not observed in everyday life. For example, if $m = 1$ kg and $L = 1$ m, then $E \geq (5.5 \times 10^{-68}$ J$)$ and the minimum energy state has a velocity of less than 10^{-33} m $\cdot$ s^{-1}. However, as we will show in Chapter 8, we live in a colorful world largely because these effects can be substantial for light particles (such as electrons) confined to small regions such as chemical bonds.

2. *Electrons cannot have planar orbits in atoms, because the position uncertainty out of the plane is then zero.* This means that the uncertainty in the momentum would be infinite, which is only possible if the length of the momentum vector (and the kinetic energy, since $K = |\vec{p}|^2/2m$) are infinite.

3. *Electron cannot simply orbit at a well-defined distance from the nucleus, even if the orbit is not planar.* In that case we could switch to spherical coordinates, and the uncertainty in the coordinate r would be zero. This would give an infinitely large uncertainty in a momentum component, and lead to the same problem as above.

4. *We can now understand why a hydrogen atom does not collapse, the way a classical electron orbit would.* Suppose we took the electronic distribution about the nucleus, and cut all distances in half.

   - The average potential energy $U = -e^2/4\pi\varepsilon_0 r$ would *double*. Since the potential energy is negative this would tend to decrease the total energy.

- Halving the orbital radius halves the position uncertainty, and the Uncertainly Principle implies that the momentum uncertainty would double. Remember that the uncertainty comes from the fact that momentum is a *vector* quantity. At any instant in time, the electron might be moving to the left or the right, giving a range of values around $|\vec{p}| = 0$. Thus doubling the uncertainty implies that $|\vec{p}|$ doubles as well.

- Since $K = |\vec{p}|^2/2m$ the minimum possible average kinetic energy *quadruples*, which would tend to increase the total energy.

Thus minimizing the total energy (potential plus kinetic) involves a tradeoff. Below a certain separation, the total energy must start to increase with further size reductions: the kinetic energy will increase more than the potential energy will decrease.

5.10 ANGULAR MOMENTUM AND QUANTIZATION OF MEASUREMENTS

In general, stable (stationary) states in quantum mechanical systems are described by a set of **quantum numbers** which give the values of all of the constants of the motion. In a hydrogen atom, for example, a classical treatment (Section 5.4) showed that the conserved quantities for a general orbit were the total energy, the length of the orbital angular momentum vector, and the direction of that vector. In a modern quantum mechanical treatment, the stationary states (called orbitals) are described by a **principal quantum number** n which gives the overall energy, just as in Equation 5.13,

$$E_n = -\frac{2.18 \times 10^{-18} \text{ J}}{n^2} \tag{5.47}$$

and by additional quantum numbers which describe either the orbital angular momentum L of the electron going around the protons (analogous to the Earth rotating around the Sun), or the intrinsic angular momentum S of the electron itself. The angular momentum of the electron is commonly called **spin**, by analogy with a spinning ball (or the Earth spinning on its own axis).

Even though angular momentum does not affect the energy (for a hydrogen atom), it certainly does play an important role in understanding bonding. Angular momentum is quantized, just as Bohr and deBroglie predicted (although the equation they derived, $|\vec{L}| = n\hbar$, is not correct). As we discuss in Chapter 6, the orbitals also have additional properties which do not correspond at all to what would be predicted classically.

Rather than detail these effects here, however, we can illustrate the effects of angular momentum in a simpler case by examining the electronic spin angular momentum $\vec{S}$. Experimentally, it is found that all electrons have angular momentum, and the length of the angular momentum vector is always $|\vec{S}| = \sqrt{3}\hbar/2$. This angular momentum generates a magnetic dipole $\vec{\mu} = \gamma \vec{S}$, where γ is a quantity called the **gyromagnetic ratio**. Thus electrons act like little bar magnets, but with some very strange properties as we discuss below.

Assume the particle is placed into an external magnetic field pointing along the z-direction ($\vec{B} = B_0 \hat{z}$). The potential energy for a magnetic dipole in such a field is:

$$U = -\mu_z B_0 = -\gamma S_z B_0 \tag{5.48}$$

The force on the dipole is:

$$F = -\frac{dU}{dz} = \gamma S_z \frac{dB_0}{dz} \tag{5.49}$$

Thus electrons can be deflected by a nonuniform magnetic field ($dB_0/dz \neq 0$). The force, hence the amount of the deflection, is proportional to the component S_z of the dipole moment along that axis.

Since all directions in space are equivalent, there is no reason to expect $\vec{S}$ to preferentially point in any specific direction. Thus we would classically expect a beam of particles with angular momentum to be deflected over a continuous range of angles, since S_z should range from $+\sqrt{3}\hbar/2$ to $-\sqrt{3}\hbar/2$. (Figure 5.14.).

Instead, electrons are deflected into only the two directions corresponding to $S_z = (\pm 1/2)\hbar$, *independent of the direction in space we choose for z.* (Figure 5.15). We refer to electrons as "spin-1/2" particles because of this property. Protons are also spin-1/2 particles, with exactly the same spin properties except for a smaller value of the gyromagnetic ratio γ.

The state $S_z = +\hbar/2$ is commonly called the "spin-up" state and $S_z = -\hbar/2$ is called the "spin-down" state, but they do not correspond to the angular momentum vector does not point exactly along the z-axis. Since $\left|\vec{S}\right|^2 = (3/4)\hbar^2 = S_x^2 + S_y^2 + S_z^2$, we

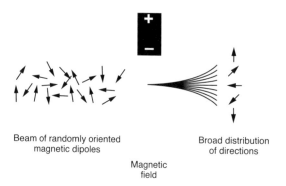

Beam of randomly oriented
magnetic dipoles

Magnetic
field

Broad distribution
of directions

FIGURE 5.14 ▶ The force on a magnetic dipole in a nonuniform field depends on the orientation. Thus a beam of randomly oriented dipoles should fan out in a wide range of directions.

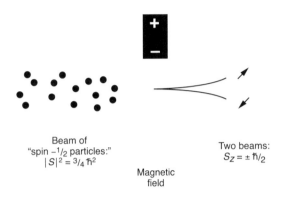

FIGURE 5.15 ▶ A beam of electrons, protons, or other spin-1/2 particles only deflects into two directions, not the continuous distribution predicted classically.

have

$$S_x^2 + S_y^2 = \frac{3}{4}\hbar^2 - S_z^2 = \frac{3}{4}\hbar^2 - \left(\frac{\pm\hbar}{2}\right)^2 = \frac{\hbar^2}{2} \qquad (5.50)$$

On average, we would expect $\left|\vec{S}_x\right|^2$ and $\left|\vec{S}_y\right|^2$ to each be $\hbar^2/4$. In fact, *every time* you measure either S_x or S_y (with a nonuniform magnetic field in the x- or y-direction instead of the z-direction) you get $\pm\hbar/2$, just as every measurement of S_z gave $\pm\hbar/2$.

Suppose we try to completely specify the direction of the angular momentum vector, for example by selecting only those particles with $S_z = S_y = S_x = +\hbar/2$. You would think you could start by selecting out the atoms with $S_z = +\hbar/2$, since the apparatus in Figure 5.15 separates them spatially from the ones with $S_z = -\hbar/2$. Then you could take these selected spins and separate them into the different possible values of S_x, using a magnetic field in the x-direction; finally, you could separate them into the different S_y values using a magnetic field in the y-direction.

Instead, if you measure S_z again after measuring S_x, you find that S_z has been randomized and $S_z = -\hbar/2$ is just as likely as $S_z = +\hbar/2$ (Figure 5.16). Only one component of the angular momentum can be specified at a time, and the act of measuring this component completely randomizes the others—just as measurements of position and momentum were limited by the Heisenberg Uncertainty Principle.

5.11 MAGNETIC RESONANCE SPECTROSCOPY AND IMAGING

Spin behavior is not just a bizarre curiosity of quantum mechanics. The difference in energy between electron spin states or nuclear spin states in a magnetic field has proven invaluable for chemistry. The largest commercially available superconducting magnets can give fields of about 20 Tesla (400,000 times stronger than the Earth's field) which are uniform to better than one part per billion. In such fields, the two states

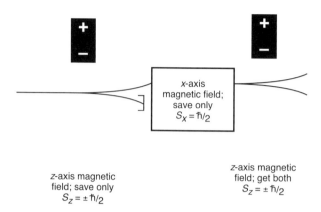

FIGURE 5.16 ▶ It is impossible to specify more than one component of the spin angular momentum vector. Measurement of any one component scrambles all of the others.

$S_z = \pm 1/2$ of a hydrogen nucleus separate in energy by a small amount, according to Equation 5.48. This separation corresponds to photon frequencies $\nu = (\Delta E)/h$ of 42 MHz per Tesla (840 MHz in a 20 Tesla magnet), in the radiofrequency range. Thus radiofrequency radiation can be absorbed by the hydrogen atoms. The value of γ for electrons is about 650 times greater than the value for hydrogen, so electron spin transitions are in the microwave range. Only molecules with an odd number of electrons or some unpaired electrons give electron spin transitions, but every hydrogen atom has nuclear spin transitions.

Not all hydrogen absorb energy at exactly the same frequency. The nuclei are sensitive to small local variations in the magnetic field. These variations arise largely from electrons in the molecule, which also act like small magnets to partially shield the proton from the external magnetic field. The strength of this electronic shielding depends on the local charge distribution. For example, a proton next to a carbonyl group ($C = O$) will absorb energy at a different frequency than one which is only next to carbons and hydrogens. This effect, called the *chemical shift*, makes proton resonance frequencies vary over a range of about 10 parts per million, which is 8400 Hertz in a 20 Tesla magnet. Individual lines are quite narrow in solution (typically about 0.1 Hertz) so even very slightly different chemical environments lead to resolved spectral lines (see Figure 5.17).

In addition, the interaction between proton spins and the electronic distribution produces an additional line splitting for physically nearby but chemically inequivalent protons, called the *scalar coupling* or *J-coupling*. Consider the *nuclear magnetic resonance (NMR)* spectra of the two molecules below (Figure 5.17). The molecule 1,2-dichloroethane has all four protons equivalent by symmetry (remember that the rotation about the C-C bond is essentially unhindered, giving only a single line). The molecule 1,1-dichloroethane has one proton on a carbon with two chlorines, and three protons (a methyl group) on a carbon with no chlorines, so the chemical shifts are different.

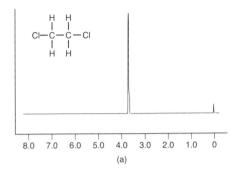

(a)

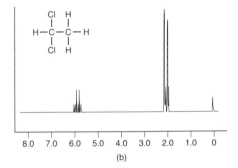

(b)

FIGURE 5.17 ▶ NMR spectra of two different molecules: 1,1-dichloroethane (top and 1,2-dichloroethane (bottom). The horizontal scale is in parts per million (PPM) away from a "reference peak" (at the far right), which by convention is dimethyl sulfoxide (DMSO). The different structure of the two spectra arises from chemical shifts and scalar couplings, and permits determination of structural features.

The methyl group protons will have transition frequencies which change slightly if the other proton is up or down. The Boltzmann distribution shows that the spin-up and spin-down states are almost equally likely for NMR at room temperature (Problem 5-13), so the methyl group is split into a doublet with equal intensity. The other proton sees eight possibilities for the spin states in the methyl group (all three spin up; two spins up three different ways; one spin up three different ways; no spins up). This breaks up the line into a 1:3:3:1 quartet with intensities given by the binomial distribution, just as we discussed in Chapter 4. An NMR spectroscopist would look at the spectrum at the bottom of Figure 5.17 and instantly realize that the lone proton had three neighbors, and that the doublet (with three times the total area of the quartet) represents three protons with one neighbor. The net effect is that the radiofrequency spectrum of any molecule is a *very sensitive indicator of molecular structure*.

Physicists pioneered NMR spectroscopy in the 1940s, and used it to understand the bizarre properties of nuclear spin angular momentum. The discovery of chemical shifts and scalar coupling made it universally applicable to chemistry, and every major chemistry department in the country has NMR spectrometers which use this effect. Many have electron spin resonance (ESR) spectrometers as well. The magnetic resonance spectra of other nuclei, such as ^{13}C and ^{31}P, can provide complementary information to the proton magnetic resonance spectrum and are also widely used. In fact, NMR is widely recognized as the single most powerful tool for determining the structures of unknown molecules. Its importance has been recognized with Nobel Prizes in physics, and more recently (1991) in chemistry to Richard Ernst. Ernst's prize was awarded largely for a technique called *two-dimensional spectroscopy*. This technique allows chemists and molecular biologists to routinely determine the full structure in solution of proteins with molecular weights greater than 25,000 grams per mole!

The magnetic resonance frequency for protons is proportional to magnetic field strength. The protons in *you* are primarily in one molecule (water) so in a uniform magnetic field, your NMR spectrum consists essentially of one line. Now suppose the magnetic field is not uniform—for example, suppose there is a magnetic field gradient of 1 mT per meter in the direction from your head to your feet. Two water molecules separated by 1 cm will see a magnetic field which differs by 10 μT. Hence they will have a resonance frequency difference of 420 Hz, much more than the NMR linewidth for either molecule (the spins will also feel a force, as shown in Equation 5.48, but since they are trapped in your body the effect of the force is negligible).

A pulse with length Δt will only excite a range of frequencies of order $\Delta \nu \approx (1/4\Delta t)$, in accord with the Uncertainty Principle. So a sufficiently long pulse would only excite the water molecules in a small region. In this case, a 1 ms pulse would excite molecules in a band about 6 mm wide. Finally, gradients in the other two spatial directions would make water molecules in different positions have different frequencies, and the NMR spectrum gives a *spatially resolved image*. This is called **magnetic resonance imaging**, or MRI; physicians dropped the term "nuclear" to avoid scaring patients in hospitals.

MRI uses only radiowaves, which are not ionizing like X-rays. It is also superior to X-rays in measuring variations in soft tissue. Figure 5.18 shows a clinical MRI brain scan (eyes and nose at top, back of head at bottom) taken with a 1.5 T magnet. The contrast in this image comes from differences in the local environment of the water molecules. For example, the water density is very high in the eyes, and the water molecules can move freely. In the brain itself the water motion is more restricted. This image is taken with a sequence of radiofrequency pulses which is designed to give less signal from the spins with restricted motion.

One of the most remarkable new developments in magnetic resonance, and indeed one of the exciting frontiers in modern clinical science, is *functional MRI* —literally watching thought processes in action. Figure 5.19 shows "fast scans" (taken in less than one second, compared to several minutes for Figure 5.18) which highlight regions in the brain which change signal strength during different cognitive tasks.

Functional imaging signals are observed by exciting the water spins with radiofrequency pulses, then waiting a long time (many milliseconds) before observing the MRI signal. The currently accepted explanation goes all the way back to basic physics.

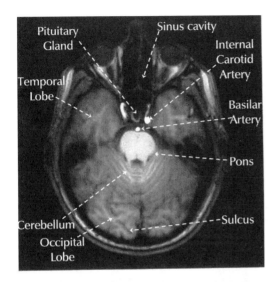

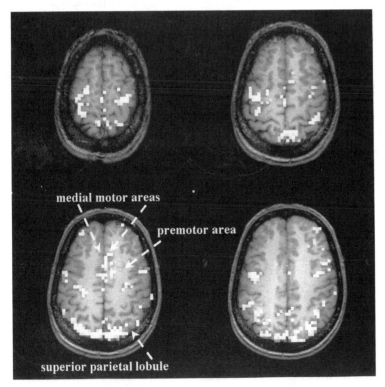

FIGURE 5.18 ▶ MRI brain scan of a healthy human, with various positions marked. The patient is lying in a 1.5 Tesla magnet. Contrast in the image comes from variations in water properties (density, linewidth) in the various tissue regions.

FIGURE 5.19 ▶ Functional magnetic resonance imaging (fMRI) detects signal changes in different regions of the brain during thought-activated processes. The four brain slices show regions which change during a complex thought process (mentally rotating two objects to see if they are superimposable or mirror images).

Oxygenated tissue and deoxygenated tissue have different water resonance frequencies, because hemoglobin (the molecule which carries oxygen) has a different number of unpaired electrons when it is oxygenated. These unpaired electrons act like bar magnets, changing the local magnetic field. Activation of the motor cortex or visual area requires oxygen transport across the blood-brain barrier, which makes the resonance frequency more uniform (decreasing the linewidth Δv). This, in turn, increases the length of time Δt before the MRI signal disappears in the activated region.

Over the next few years, functional imaging has the potential to completely revolutionize our understanding of the mind. Twentieth-century chemistry and physics transformed biology into "molecular biology"; enzymes and proteins became understood as big molecules instead of black boxes, and the molecular basis of life processes was developed. Scientists are now on the threshold of acquiring this same level of understanding of processes in cells and organs.

5.12 SUMMARY

In this chapter we have tried to give an overview of the critical experiments which proved that classical mechanics and electromagnetic theory are only valid in special cases, and that the range of validity does not include atoms and molecules. We have presented a number of very strange results, all of which have been amply verified by experiment; and we have summarized the work of twelve Nobel Prize winners in only a few pages. But this is far from the end of the story. Applications of quantum mechanics are still the subject of vigorous ongoing research, and the philosophical consequences are much too subtle to explore in the limited space available here. Still, a few illustrations of some of the stranger applications might be useful for perspective.

Quantum mechanics provides an interesting method for communicating data. Particles with $S_z = +\hbar/2$ are commonly called "spin up" and $S_z = -\hbar/2$ are "spin down," let us call $S_x = +\hbar/2$ the "spin right" state, and $S_x = -\hbar/2$ the "spin left" state. Now suppose we design a communications system which sends our a signal as a stream of bits (ones and zeroes). Suppose the person transmitting the data (conventionally called Alice) sends each bit on a single particle. She can use *either* spin up-spin down as "1" and "0," or spin right-spin left as "1" and "0." She makes this choice of *basis states* (up-down or left-right) at random each particle she sends. So each particle will be sent out in one of four different states (up, down, left, and right), as in Figure 5.20.

The person who receives the stream of particles (conventionally called Bob) does not know, in advance, whether Alice chose up-down or right-left to encode the information on any given particle. He can decode the stream of data only by guessing up-down or right-left, because measurement of S_x randomizes S_z (and *vice versa*) and S_z and S_x cannot both be measured on the same particle (Figure 5.21). Roughly half the time he will guess correctly, and get the correct "1" or "0." Roughly half the time he will guess incorrectly, in which case he will get either "1" or "0" at random—this half of the data will be garbage.

Data	1	0	1	1	0	0	...
Basis (random)	UD	RL	RL	UD	UD	RL	
Transmitted particle	↑	←	→	↑	↓	←	

FIGURE 5.20 ► Alice transmits a data stream by encoding each data bit either as $1 =$ spin up and $0 =$ spin down, or as $1 =$ spin right and $0 =$ spin left. She chooses whether to use up-down or right-left at random, and chooses again for each data bit. Each bit is then sent as a single particle (for example, one electron).

Alice and Bob then tell each other what basis states they used for each particle (not the data itself), and the incorrect data is resent by the same procedure. Again roughly half of the data gets through, and roughly half is lost; eventually, repeating this procedure many times, all of the data can be transmitted.

This approach is called *quantum cryptography*, and systems have been demonstrated which have enough bandwidth to transmit speech (practical systems use photons instead of particles, but the principle is identical). Why go to all this bother? Because the message *cannot be eavesdropped without the sender and receiver knowing something is wrong*. In conventional communications, someone else can intercept the transmitted signal and retransmit it without alteration, or can tap off only a small part of the signal. But here the message is sent one particle at a time, so it is not possible to tap off a small portion. In addition, the eavesdropper (conventionally called Eve) cannot know, when a particle is received, if it was encoded up-down or left-right. So she cannot detect the signal then send a duplicate. Any attempt to eavesdrop on the signal will reduce it to garbage, for both the receiver and the eavesdropper.

Another example of current interest is *quantum computing*. Conventional computer systems have improved exponentially over the last four decades, roughly doubling processor speed, memory size, and disk size every eighteen months. This kind of improvement has required very expensive research and development (billions of dollars in manufacturing cost for the fabrication lines which make each new generation of chips), but massive computer sales have offset this investment. The exponential growth cannot go on forever. Even now, the smallest logic gates on the fastest chips switch

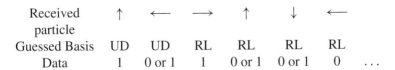

Received particle	↑	←	→	↑	↓	←	
Guessed Basis	UD	UD	RL	RL	RL	RL	
Data	1	0 or 1	1	0 or 1	0 or 1	0	...

FIGURE 5.21 ► In order to decode the received data stream, the receiver has to guess whether the transmitter used up-down or right-left to encode the particles. If the guess is correct, the received data is right. If the guess is wrong, the received data will be "0" half the time, and "1" half the time at random.

states using fewer than 100 electrons. By 2015, if we simply follow historical trends, the gates must switch with half an electron, and the cost of building the microprocessor fabrication line will exceed the sum of the gross national products of the planet.

However, since one electron or proton has two possible spin states, N electrons or protons have 2^N possible states (just like the coin toss case). Molecules can be configured to be simultaneously in many different quantum states, just as the electron in the two-slit experiment seems to pass through both slits simultaneously. In principle, this property can be used someday to make massively parallel computers, and such computers with five or six bits have been made in the laboratory (using NMR). As of this writing, nobody knows whether or not it will ever be possible to build a quantum computer which is big enough to do a computation faster than a conventional machine, although it is clear that NMR will not work for this application.

We conclude this chapter by going back to Albert Einstein, whose work was instrumental in the evolution of the quantum theory. Einstein was unable to tolerate the limitations on classical determinism that seem to be an inevitable consequence of the developments outlined in this chapter, and he worked for many years to construct paradoxes which would overthrow it. For example, quantum mechanics predicts that measurement of the state of a system at one position changes the state everywhere else immediately. Thus the change propagates faster than the speed of light—in violation of at least the spirit of relativity. Only in the last few years has it been possible to do the appropriate experiments to test this ***ERP paradox*** (named for Einstein, Rosen and Podolsky, the authors of the paper which proposed it). The predictions of quantum mechanics turn out to be correct.

If you do not understand quantum theory completely, you are in good company. For a fuller treatment of the philosophical consequences of quantum mechanics see reference [4].

For more information on the history of the development of quantum theory see reference [5].

▶ PROBLEMS ▶

5-1.* The most common oxide of a certain metal M contains 1 kg of oxygen for each 2.33 kg of the metal.

(a) Find the atomic weight of M if the oxide is assumed to be
i) M_2O, ii) MO, iii) M_2O_3, iv) MO_2, v) M_2O_5.

(b) Heat capacities are more difficult to measure accuracy than combining ratios. An experimental value of c_v for the metal gives $0.42 \pm .04$ kJ $\cdot$ kg^{-1} $\cdot$ K^{-1}. Using these two results, what is the formula of the oxide and the atomic weight of the metal?

5-2. Show that Equation 5.9, the Rayleigh-Jeans law, is identical to the Planck blackbody distribution in the limit as $h \to 0$.

5-3. Explain how infrared detectors can be used as "night vision goggles" to distinguish between a warm engine and the surrounding grass, even on a cloudy night.

5-4. For $\lambda = 500$ nm (near the peak sensitivity of human vision), find the ratio between the amount of light produced by a 2600K lightbulb filament (1 mm wide, 1 cm long), and the amount of light produced by a 800K stove burner (area 0.1 m^2). (This explains why a stove can be hot enough to burn you badly even though you cannot see it is hot.)

5-5. With some algebraic manipulation, the correct expression for the average energy of an oscillator which is restricted to energies $E = 0, h\nu, 2h\nu, \ldots, nh\nu, \ldots$ can be shown to be:

$$\langle E \rangle = \frac{h\nu}{e^{h\nu/k_B T} - 1}$$

Use this formula to calculate the heat capacity of an oscillator. What is the heat capacity when $h\nu = 100 k_B T$? Also, use this equation to find the heat capacity in the limit $h\nu \ll k_B T$.

5-6. The "dot product" of two vectors $\vec{u}$ and $\vec{v}$, written $\vec{u} \cdot \vec{v}$, is a number given by the expression $\vec{u} \cdot \vec{v} = u_x v_x + u_y v_y + u_z v_z = |\vec{u}| |\vec{v}| \cos \theta$, where θ is the angle between the two vectors. Use this equation to show that $\vec{u}$ and $\vec{v}$ are perpendicular for a circular orbit.

5-7. A mole of photons is given the name "einstein." A typical red laser pointer produces 5 mW average power at a wavelength $\lambda = 650$ nm. How long does it take this pointer to produce one einstein?

5-8. The Earth (radius 6378 km) has an approximately circular orbit of radius 1.496×10^8 km. It goes around the sun once a year (3.2×10^7 s). It rotates on its own axis once a day (84,400 s). Its mass is 5.98×10^{24} kg; the Sun's mass is 1.989×10^{30} kg.

(a) Calculate the angular velocity ω and the mean rotational speed $|\vec{v}|$ for the Earth's rotation about the Sun.

(b) Calculate the reduced mass μ, the moment of inertial I, the angular momentum $|\vec{L}|$ and the rotational kinetic energy $K = |\vec{L}|^2 /2I$ for the Earth's motion around the Sun. You may assume the orbit is circular.

(c) If you consider only the rotation of the Earth, how much acceleration are you experiencing if you sit in a "motionless" chair at the equator?

5-9. The neutron persists outside of a nucleus for approximately 12 minutes before decaying. Use the uncertainty principle to estimate the fundamental limitation to measurements of its mass.

5-10. The expressions for the Bohr atom technically should use the reduced mass $\mu = m_{\text{electron}} m_{\text{nucleus}} / (m_{\text{electron}} + m_{\text{nucleus}})$ instead of the electron mass, as noted in Equation 5.22. This alters the calculated value of the Bohr radius a_0, and therefore also alters the radius R and the total energy $E = -Ze^2/2R$.

(a) Calculate the reduced mass $\mu = m_1 m_2/(m_1 + m_2)$ of a hydrogen atom and compare it to the mass of the electron alone.

(b) Deuterium is an isotope of hydrogen with a proton and a neutron in the nucleus, instead of just a proton. This changes the reduced mass. Find the change in the frequency of the $n = 2$ to $n = 1$ emission line in going from hydrogen to deuterium (this is easily measured in the laboratory).

5-11. (a) Use the Bohr model of the hydrogen atom to calculate the kinetic energy of an electron in the $n = 1$ state.

(b) The kinetic energy gives us the expected length of the momentum vector ($K = |\vec{p}|^2/2m$), but the direction of the momentum vector is random. Use this result to estimate Δp, the uncertainty in the momentum vector.

(c) Now use the Heisenberg uncertainty relationship $\Delta x \Delta p \approx h/4$ to *estimate* the position uncertainty for a 1 s state. Compare this to the expected radius a_0 (Equation 5.16).

5-12. The shortest laser pulse created to date has a duration (full width at half maximum) of 3.5 femtoseconds, and a center wavelength of approximately 800 nm ($\nu \approx 375$ THz). However, because of the uncertainty principle, such a pulse has a very large range of frequencies $\Delta \nu$. Use the uncertainty principle to determine $\Delta \nu$.

5-13. In zero magnetic field, the two spin states of a proton (the spin-up state and the spin-down state) have the same energy. In a large magnetic field (10 Tesla), these two states are separated in energy (see Equation 5.48). The spins can be "flipped" by radiation which has exactly the right energy per photon to promote the protons from the ground state to the excited state.

(a) The gyromagnetic ratio γ for protons is 2.6752×10^8 rad $\cdot$ s^{-1} $\cdot$ (Tesla)$^{-1}$. Verify that 426 MHz radiowaves give photons with the same energy as the splitting between the two levels at 10 Tesla.

(b)* Spin energy differences are very small compared to $k_B T$ near room temperature. One consequence of this is a relatively large population in the more excited state. Use the Boltzmann distribution to calculate the fraction of the population in the higher state at 300K.

(c) The net magnetization M of the entire sample is proportional to the difference in population between the spin-up state and the spin-down state. Use the Taylor expansion of e^x to show that, near room temperature, the net magnetization is proportional to the reciprocal of the temperature ($M \propto 1/T$).

5-14. Lasers can be used to essentially stop moving atoms. In a typical application, sodium atoms absorb light with wavelength $\lambda \approx 590$ nm, thus decreasing their velocity in the direction of the laser beam. After the light is emitted in a random direction, the atom is free to absorb again.

Use conservation of momentum to determine how much the velocity in the direction of the laser beam changes by the absorption of one photon.

5-15. The de Broglie relationship can be used to predict the possible energies for a particle in a one dimensional "box". Since the walls of the box are infinitely high, assume that the box itself must contain an integral or half-integral number of wavelengths, with the zeroes of the wave at either edge ($\lambda = 2L/n$, n a positive integer). Use this relation to derive an expression for the energy as a function of n.

5-16. A photomultiplier tube is designed so that it gives a short burst of current ($\approx 10^6$ electrons in 10 ns) when a photon hits it, as discussed in Section 5.3. Since it is an electronic device, it also has noise.

(a) Assume that the only source of noise has a Gaussian distribution. Assume that the average value is 0 mA, the standard deviation is 4 μA, and assume that the noise can be completely different every 10 ns (so that, in effect, the noise has 100,000,000 independent chances per second to be large enough to fool you into thinking an electron hit). In one second, how many times out of those 100,000,000 will the noise exceed 10^6 electrons in 10 ns, thus making you think you have detected a photon even if the tube is dark (this is called the "dark current")?

(b) Suppose the circuit is improved slightly, so that now the standard deviation of the noise is 2 μA instead of 4 μA. How much does the dark current decrease?

Chapter 6

Applications of Quantum Mechanics

> Thus, the task is, not so much to see what no one has yet seen; but to think what nobody has yet thought, about that which everyone sees.
>
> *Erwin Schrödinger* (1886–1961)

6.1 WAVE MECHANICS

By 1925 the hodgepodge of bizarre results in the last chapter had led to the complete collapse of classical physics. What was now needed was some framework to take its place. Werner Heisenberg and Erwin Schrödinger came up with two apparently very different theoretical descriptions; within a year, however, it had become clear that both approaches were in fact identical, and they still stand as the foundations of modern quantum mechanics.

Schrödinger's description, called *wave mechanics*, is the easier one to present at the level of this book. A general description requires multivariate calculus, but some useful special cases (such as motion of a particle in one dimension) can be described by a single position x, and we will restrict our quantitative discussion to these cases.

6.1.1 Prelude—Imaginary and Complex Numbers

Before we go into the mathematical framework behind wave mechanics, we will review one more mathematical concept normally seen in high school: imaginary and complex numbers. As discussed in Section 1.2, for a general quadratic equation $ax^2 + bx + c =$

0, $x = (-b \pm \sqrt{b^2 - 4ac})/2a$ gives two solutions if $b^2 > 4ac$ and one solution if $b^2 = 4ac$. If $b^2 < 4ac$, no real numbers are solutions to the quadratic equation. For example, the equation $x^2 + 1 = 0$ has no real solutions.

If we define $i = \sqrt{-1}$ (an ***imaginary*** number) then we can still write out two solutions for $b^2 < 4ac$. In general these solutions will have both a real and an imaginary part, and are called ***complex numbers***. For example, if $x^2 - 2x + 2 = 0$, the solutions are $x = 1 \pm i$. Complex solutions to quadratic equations are not important in chemical problems; however, complex numbers themselves will prove to be important in quantum mechanics.

We can write a general complex number in the form $z = x + iy$, and graph such numbers in an xy-plane (the ***complex plane***, see the left side of Figure 6.1). The distance from the origin $|z| = \sqrt{x^2 + y^2}$ is called the ***magnitude*** of the complex number; it is a generalization of the concept of absolute value for real numbers. The angle θ is called the *phase* of the complex number. Purely real numbers have $\theta = 0$ or π; purely imaginary numbers have $\theta = \pi/2$ or $3\pi/2$. Phases between 0 and 2π are sufficient to describe any number. The number $x - iy$ (reversing the sign of only the imaginary part) is called the *complex conjugate* of z, and is written z^*. Note that $zz^* = |z|^2$.

The Taylor series expansion in Chapter 2 makes it possible to derive a remarkable relationship between exponentials and trigonometric functions, first found by Euler:

$$e^{i\theta} = \cos\theta + i\sin\theta \tag{6.1}$$

You should convince yourself that $\left|e^{i\theta}\right| = 1$, and that the phase of $e^{i\theta}$ is just θ, as shown on the right-hand side of Figure 6.1.

All of the usual properties of exponentials (Equations 1.13 and 1.14) also apply to complex exponentials. For example, the product of two exponentials is found by sum-

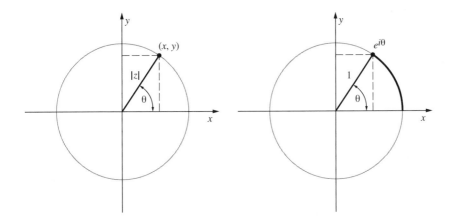

FIGURE 6.1 ▶ **Left**: representation of $z = x + iy$ in the "complex plane," showing the magnitude $|z|$ and phase θ. **Right**: the quantity $e^{i\theta}$ is always on the unit circle, and is counterclockwise by an angle θ from the x-axis.

ming the exponents: $e^{1+2i} \cdot e^{3+4i} = e^{(1+3)+(2+4)i} = e^{4+6i}$. The magnitude of this complex exponential is determined completely by the real term in the exponent:

$$\left| e^{4+6i} \right| = \left| e^4 \right| \cdot \left| e^{6i} \right| = \left| e^4 \right| \cdot 1 = e^4$$

The phase (6 radians) is determined completely by the imaginary term in the exponent. The relations $\sin(-\theta) = -\sin\theta$ and $\cos(-\theta) = \cos\theta$ imply

$$
\begin{aligned}
e^{-i\theta} &= e^{i(-\theta)} = \cos(-\theta) + i\sin(-\theta) \\
&= \cos\theta - i\sin\theta
\end{aligned}
\tag{6.2}
$$

Finally, Equations 6.1 and 6.2 can be combined to give

$$\cos\theta = (e^{i\theta} + e^{-i\theta})/2; \quad \sin\theta = (e^{i\theta} - e^{-i\theta})/2i \tag{6.3}$$

6.1.2 Wavefunctions and Expectation Values

Schrödinger's picture of quantum mechanics describe any object (for example, an electron) by its **wavefunction**, $\psi(x)$. The wavefunction itself is not directly observable, but it contains information about all possible observations because of the following two properties.

1. $\psi^*(x)\psi(x) = |\psi(x)|^2 = P(x)$, the probability of finding the object at position x. $P(x) \geq 0$ by definition, but the wavefunction is not just the square root of the probability. Wavefunctions at any point can be positive, negative, or even complex. This **phase variation** of the wavefunction is central to quantum mechanics. It lets particles exhibit wave-like behavior such as interference at positions where two waves are out of phase, just as classical waves exhibit interference (Chapter 3).

   Since the object must be *somewhere*, with probability 1,

   $$\int_{x=-\infty}^{+\infty} |\psi(x)|^2 \, dx = \int_{x=-\infty}^{+\infty} P(x) \, dx = 1 \tag{6.4}$$

   All possible wavefunctions are continuous (no breaks or jumps) and satisfy Equation 6.4.

2. Given any observable quantity A (for example, the position or momentum of the object), the wavefunction $\psi(x)$ lets us calculate the **expectation value** $\langle A \rangle$ which is the average value you would get if you made a very large number of observations of that quantity. The wavefunction thus contains all the information which can be predicted about the system.

For some observables, such as the position x or the potential energy U, the expectation value is calculated from the probability distribution in exactly the same way as we calculated classical averages in Chapter 4:

$$\langle x \rangle = \int_{x=-\infty}^{+\infty} x P(x)\, dx = \int_{x=-\infty}^{+\infty} x\, |\psi(x)|^2\, dx;$$

$$\langle U \rangle = \int_{x=-\infty}^{+\infty} U(x) P(x)\, dx = \int_{x=-\infty}^{+\infty} U(x)\, |\psi(x)|^2\, dx \qquad (6.5)$$

Some expectation values require more complicated integrals. There is a general prescription for determining the correct integral, but here we will merely give the right form for two other important quantities: the momentum $\langle p \rangle$ and the kinetic energy $\langle K \rangle$:

$$\langle p \rangle = i\hbar \int_{x=-\infty}^{+\infty} \psi^*(x)\frac{d\psi(x)}{dx}\, dx \qquad (6.6)$$

$$\langle K \rangle = -\frac{\hbar^2}{2m} \int_{x=-\infty}^{+\infty} \psi^*(x)\frac{d^2\psi(x)}{dx^2}\, dx \qquad (6.7)$$

Notice that these equations explicitly include derivatives and the complex conjugate ψ^* of the wavefunction. The expression for the momentum even includes $i = \sqrt{-1}$! **Complex numbers are not just a mathematical convenience in quantum mechanics; they are central to the treatment.** Equation 6.6 illustrates this point directly. Any measurement of the momentum (for example, by measuring velocity and mass) will of course always give a real number. But if the wavefunction is purely real, the integral on the right-hand side of Equation 6.6 is a real number, so the momentum is a real number multiplied by $i\hbar$. The only way that product can be real is if the integral vanishes. Thus *any real wavefunction corresponds to motion with no net momentum.* Any particle with net momentum must have a complex wavefunction.

6.1.3 Schrödinger's Equation and Stationary States

If the wavefunction satisfies *Schrödinger's equation*,

$$-\frac{\hbar^2}{2m}\frac{d^2\psi}{dx^2} + U(x)\psi(x) = E\psi(x) \qquad (6.8)$$

with the same value of E for every value of x, then we call ψ a *stationary state* with total energy E (potential plus kinetic). Sometimes a stationary state is also called an *eigenstate*.

You have very likely seen stationary states before, although the name might be new to you. The orbitals in a hydrogen atom ($1s$, $2p_z$, and so forth) are all stationary states, as we will discuss in Section 6.3. The probability distribution $P(x)$ and the expectation values of all observables are constant in time for stationary states.

If ψ is a stationary state, then we can multiply Schrödinger's equation by -1, and show that $-\psi$ is also a stationary state:

$$-\frac{\hbar^2}{2m}\frac{d^2(-\psi(x))}{dx^2} + U(x)\,(-\psi(x)) = E\,(-\psi(x)) \tag{6.9}$$

The probability density $P(x) = |\psi(x)|^2$ is the same for ψ as it is for $-\psi$; the expectation values for all observable operators are the same as well. In fact, we can even multiply ψ by a complex number and the same result holds. The overall phase of the wavefunction is arbitrary, in the same sense that the zero of potential energy is arbitrary. Phase *differences* at different points in the wavefunction, on the other hand, have very important consequences as we will discuss shortly.

Equation 6.8 does not always have to be satisfied; $\psi(x)$ does not have to be a stationary state. However, if $\psi(x)$ does not satisfy Equation 6.8, the probability distribution $P(x)$ and the expectation values of observables will change with time. The stationary states of a system constitute a *complete basis set*—which just means that any wavefunction ψ can be written as a superposition of the stationary states:

$$\psi(t=0) = a_1\psi_1 + a_2\psi_2 + a_3\psi_3 \cdots = \sum_i a_i\psi_i \tag{6.10}$$

Note that the label "(x)" has been dropped in Equation 6.10, and in many other equations in this chapter. It is still understood that the wavefunction depends on position, but eliminating the label simplifies the notation.

If the wavefunction at time $t = 0$ is given by Equation 6.10, it can be shown that the wavefunction at any later time is given by:

$$\psi(t) = a_1 e^{-iE_1 t/\hbar}\psi_1 + a_2 e^{-iE_2 t/\hbar}\psi_2 + a_3 e^{-iE_3 t/\hbar} \cdots = \sum_i a_i e^{-iE_i t/\hbar}\psi_i \tag{6.11}$$

Equation 6.11 shows that even if a wavefunction is initially real, it later becomes complex. *For more information on wave mechanics*: see Reference [4]

6.2   PARTICLE-IN-A-BOX: EXACT SOLUTION

The "particle-in-a-box" problem, which we considered qualitatively in Chapter 5, turns out to be one of the very few cases in which Schrödinger's equation can be exactly solved. For almost all realistic atomic and molecular potentials, chemists and physicists have to rely on approximate solutions of Equation 6.8 generated by complex computer programs. The known exact solutions are extremely valuable because of the *insight* they provide—they let us make predictions about the properties of the solutions in more complicated cases.

Schrödinger's equation contains the product $U(x)\psi(x)$. Recall from Chapter 5 that the "particle-in-a-box" potential is infinite except inside a box which stretches from $x = 0$ to $x = L$. So the product $U(x)\psi(x)$ would be infinite for $x < 0$ *or* $x > L$ unless $\psi(x < 0) = \psi(x > L) = 0$. Thus if the energy E is finite, the wavefunction can be nonzero only for $0 < x < L$—in other words, the particle is inside the box.

Inside the box the wavefunction must satisfy the equation

$$-\frac{\hbar^2}{2m}\frac{d^2\psi}{dx^2} = E\psi(x) \tag{6.12}$$

or equivalently

$$\frac{d^2\psi}{dx^2} = -\frac{2mE}{\hbar^2}\psi(x) \tag{6.13}$$

since $U(x) = 0$. The second derivative of ψ is proportional to ψ itself, with a negative sign. But this was exactly the situation we encountered with the harmonic oscillator in Chapter 3. So the answer must be the same:

$$\psi(x) = A\sin(\alpha x) + B\cos(\alpha x) \tag{6.14}$$

Substitution of 6.14 into 6.13 gives

$$\alpha = \sqrt{\frac{2mE}{\hbar^2}}. \tag{6.15}$$

Because wavefunctions must be continuous, we also have *boundary conditions*—the wavefunction must vanish at $x = 0$ and $x = L$.

$$\psi(0) = A\sin(0) + B\cos(0) = B = 0 \tag{6.16}$$
$$\psi(L) = A\sin(\alpha L) = 0 \tag{6.17}$$

Equation 6.17 implies either that $A = 0$, in which case $\psi = 0$ everywhere and the wavefunction cannot be normalized (Equation 6.4), or

$$\alpha L = n\pi, n = 1, 2, 3\ldots \tag{6.18}$$

Thus we have

$$\psi_n(x) = A\sin(n\pi x/L),\ 0 < x < L;\ \psi_n(x) = 0\text{ otherwise} \tag{6.19}$$

The subscript on the wavefunction identifies it as the one with some particular value of n. $n = 0$ would force the wavefunction to vanish everywhere, so there would be no probability of finding the particle anywhere. Hence we are restricted to $n > 0$.

A is just the normalization constant. Substituting Equation 6.19 into Equation 6.4 gives $A = \sqrt{2/L}$ (Problem 6-5):

$$\psi_n(x) = \sqrt{2/L}\,\sin(n\pi x/L),\, 0 < x < L;\, \psi_n(x) = 0 \text{ otherwise}$$

(6.20)

The energy can be found by combining Equation 6.18 with the definition of α (Equation 6.15) and solving for E:

$$E_n = \frac{n^2 h^2}{8mL^2}$$

(6.21)

The three wavefunctions ψ_1, ψ_2 and ψ_3 are graphed in Figure 6.2.

Notice that the number of zero crossings (nodes) increases as the energy increases. This is a very general result which applies to atomic and molecular wavefunctions as well, as we discuss later.

This energy is entirely kinetic energy, since $U(x)$ vanishes in the box. Since $E = |\vec{p}|^2/2m$, we can also calculate the length of the momentum vector for the nth state:

$$|\vec{p}| = \sqrt{2mE} = \frac{nh}{2L}$$

(6.22)

Since this is a one-dimensional problem, there are only two choices: the momentum is either pointed along $+x$ or $-x$, so $p_x = \pm nh/2L$. However, for any of the stationary states (or any other real wavefunction, as discussed earlier), Equation 6.6 gives $\langle p \rangle = 0$. So a particle in any of the stationary states is not moving on average. Any single observation would give $p_x = \pm nh/2L$, but the average of many observations would be zero.

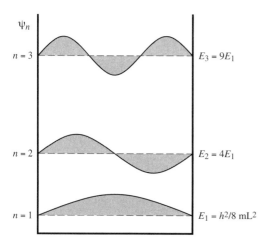

FIGURE 6.2 ▶ The three lowest energy wavefunctions for a particle in a box.

Solving Equation 6.5 gives $\langle x \rangle = L/2$, which is also obvious from inspection. All of the probability distributions for the stationary states are symmetric about the center of the box.

Of course it is possible for the particle to be someplace other than the center, or to have nonzero momentum. We make such wavefunctions with combinations of the stationary states. Figure 6.3 illustrates the wavefunction and the probability distribution obtained by adding together the two lowest solutions, then dividing by $\sqrt{2}$ to normalize the wavefunction (Problem 6-5). These superposition states are *not* stationary states because the two stationary states we added together have *different energies*. This implies that such quantities as the momentum and average position will change with time. Notice that these two stationary states constructively interfere on the left side of the box, and destructively interfere on the right side. As a result, the wavefunction is localized mainly in the left side of the box ($\langle x \rangle < L/2$). However, since the wavefunction is real, it is still true that $\langle p \rangle > 0$. If we had taken the difference instead of the sum of the two wavefunctions, the particle would be localized in the right side of the box (Problem 6-7). Notice also that the wavefunction still does not rise rapidly near the edge of the box. If we added in still higher energy wavefunctions, it would be possible to create a sharper rising edge. As noted earlier, any possible wavefunction in the box can be written as a unique combination of the stationary states.

Since the wavefunction is not a stationary state, it evolves according to Equation 6.11. If there are only two stationary states in the superposition state, as in Figure 6.3, the probability distribution and all of the observables oscillate at a frequency $\omega = (E_2 - E_1)/\hbar$ (see Problem 6-10). If we have the "left side" wavefunction in Figure 6.3 at time $t = 0$, at later times we will have a "right side" wavefunction. At

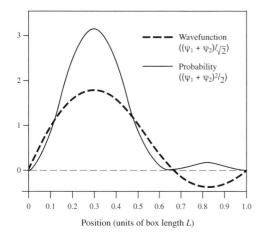

FIGURE 6.3 ▶ A wavefunction created by adding together the first two stationary states is no longer centered in the box.

intermediate times, the momentum becomes nonzero (see Problem 6-10). Note also that this oscillation frequency is the same as the frequency of a photon whose energy is equal to the energy difference between the two states. The electric field of a photon oscillates as well, which suggests that absorption of such a photon might be able to change the relative population of the two states. We will discuss this in Chapter 8.

This concept of superposition states is at the heart of many quantum mechanical problems. At the end of the last chapter we discussed the two stationary states of a spin-1/2 particle in a magnetic field oriented along the z-direction: the *spin up* state $S_z = +\hbar/2$ and the *spin down* state $S_z = -\hbar/2$. Each of these states has a nonzero value of the spin angular momentum measured along the z-axis, but zero average value of the spin angular momentum along any direction in the xy-plane. $(\langle S_x \rangle = \langle S_y \rangle = 0)$. Superpositions of these states create the spin right and spin left states discussed in the last chapter, or in general states with a nonzero value of the spin angular momentum along any specific direction of interest. For example, in more advanced courses it is shown that

"spin right" $(S_x = +\hbar/2) = \{\text{"spin up"} + \text{"spin down"}\}/2^{1/2}$
"spin left" $(S_x = -\hbar/2) = \{\text{"spin up"} - \text{"spin down"}\}/2^{1/2}$

Thus each of these states has equal amounts of "spin up" and "spin down"; they differ only by the phase in the superposition.

6.3 SCHRÖDINGER'S EQUATION FOR THE HYDROGEN ATOM

Schrödinger's equation for a single electron and a nucleus with Z protons is an extension into three dimensions (x, y, z) of Equation 6.8, with the potential replaced by the Coulomb potential $U(r) = Ze^2/4\pi\varepsilon_0 r$. This problem is exactly solvable, but it requires multivariate calculus and some very subtle mathematical manipulations which are beyond the scope of this book.

The wavefunctions for a hydrogen atom are described by the **principal quantum number n**, which gives the energy; the **orbital angular momentum quantum number l**; and the **azimuthal quantum number m_l**, which gives the z-component of orbital angular momentum. In addition, for each of these wavefunctions, we can have $m_s = \pm 1/2$, corresponding to the projection of the spin angular momentum along the z-axis.

$$
\begin{aligned}
E_n &= (-2.18 \times 10^{-18} \text{ J})Z^2/n^2 \\
\left| \vec{L} \right|^2 &= \hbar^2 l(l+1); l = 0, 1, 2, \ldots (n-1) \\
L_z &= \hbar m_l, m_l = -1, -1+1, \ldots l-1, l \\
S_z &= (\pm 1/2)\hbar
\end{aligned}
\qquad (6.23)
$$

We noted in Chapter 5 that only one component of the electron's angular momentum (typically chosen as the z-component, S_z) can be specified. A similar result holds for

L; only the single component L_z can be specified. There are $(2l + 1)$ levels of the azimuthal quantum number m_l for each value of l.

Note that even when m_l has its maximum value ($m_l = l$), $L_z^2 < |\vec{L}|^2$. Thus we never know the exact direction of the angular momentum vector. An atom does not really have any preferred direction in space, so all of the different m_l levels have exactly the same energy: such levels are called *degenerate*. In fact, for a hydrogen atom, states with different values of l, m_l, and m_s (but the same value of n) are degenerate as well.

The orbital angular momentum quantum number l has other strange characteristics. Notice that $l = 0$ is allowed. However, for a classical orbit, circular or elliptical, we have $|\vec{L}| = mv_\perp R$ (Equation 5.19), which cannot be zero. The vector L points in the direction perpendicular to the orbit, so if $l = 0$ the orbit must be equally likely to be in any direction. This suggests (and calculations confirm) that the orbitals with $l = 0$ are spherically symmetric. These are commonly called "s orbitals." The symmetry of the orbital gets progressively more complicated as we go to higher values of l:

l	0	1	2	3	4
orbital name	s	p	d	f	g

The $n = 1$ and $n = 2$ orbital wavefunctions for a one-electron atom are:

$$n = 1, l = 0 \text{ ("1s orbital"): } \psi_{1s} = C_1 \exp(-Zr/a_0)$$
$$n = 2, l = 0 \text{ ("2s orbital"): } \psi_{2s} = C_2(2 - Zr/a_0)\exp(-Zr/2a_0)$$
$$n = 2, l = 1 \text{ ("2p orbitals"): }$$

$$
\begin{aligned}
\psi_{2p_x} &= C_3 x \exp(-Zr/2a_0) = C_3[r\sin\theta\cos\phi\exp(-Zr/2a_0)] \quad &(6.24)\\
\psi_{2p_y} &= C_3 y \exp(-Zr/2a_0) = C_3[r\sin\theta\sin\phi\exp(-Zr/2a_0)]\\
\psi_{2p_z} &= C_3 z \exp(-Zr/2a_0) = C_3[r\cos\theta\exp(-Zr/2a_0)]
\end{aligned}
$$

where $r = \sqrt{x^2 + y^2 + z^2}$, Z is the number of protons in the nucleus, and $a_0 \equiv 4\pi\varepsilon_0\hbar^2/m_e e^2 = 52.92$ pm is the Bohr radius introduced in Chapter 5. The coefficients C_1, C_2, and C_3 are normalization constants, chosen to satisfy the three-dimensional version of Equation 6.4. The energy can also be written in terms of the Bohr radius:

$$E_n = \frac{-Z^2 e^2}{8\pi\varepsilon_0 a_0 n^2} \quad (6.25)$$

Every introductory chemistry textbook graphs the lowest energy solutions in a variety of different ways, and we will not duplicate those graphs here. However, some of the important characteristics of these solutions are often omitted from introductory texts.

1. There is no unambiguous way to represent $\psi(x, y, z)$ on a two-dimensional sheet of paper. The most common convention is to draw a *contour*—a surface which

consists entirely of points with the same value of $|\psi|^2$, and which encloses some large fraction (say 90%) of the total probability density $|\psi|^2$.

The contours for s wavefunctions are all spheres, since the value of ψ depends only on r (Figure 6.4). Any contour with $|\psi|^2$ for the $2p_z$ orbital completely misses the xy-plane (where $z = 0$, hence $\psi = 0$). Any contour for $2p_z$ orbitals breaks down into two lobes with opposite signs of ψ—usually represented as a positive lobe above and a negative lobe below the plane.

2. All of the hydrogen stationary states can be written in the general form:

$$\psi(r, \theta, \phi) = R_{n,l}(r)Y_{l,m_l}(\theta, \phi) \tag{6.26}$$

which separates the radial part of the wavefunction $R_{n,l}(r)$ from the angular part $Y_{l,m_l}(\theta, \phi)$. As the notation implies, the radial part depends only on the n and l quantum numbers, not on m_l, and the angular part is independent of n. In general, the radial part of the wavefunction $R_{n,l}(r)$ vanishes for $n - l - 1$ different values of r. For example, the $2s$ wavefunction ($n = 2, l = 0$) vanishes everywhere on the surface of a sphere with radius $r = 2a_0/Z$. These surfaces are called **radial nodes**.

The angular part is the same for a $2p_z$, $3p_z$, or $110p_z$ orbital, since only n changes between these wavefunctions. The angular part also makes the wavefunction vanish on l other surfaces, called *angular nodes*. For a p_z orbital, the lone angular node is a simple surface—the xy-plane. The $2p_z$ and $3p_z$ wavefunction look different in Figure 6.4 because of the extra radial node in the $3p_z$ case. For higher orbitals the angular nodes can look quite complicated.

It is also common to graph only the angular part $Y_{l,m_l}(\theta, \phi)$ of an orbital. For any p_z orbital ($l = 1, m_l = 0$) this gives two spheres.

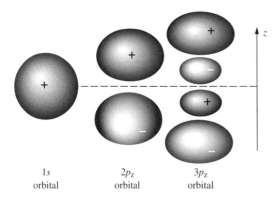

| $1s$ orbital | $2p_z$ orbital | $3p_z$ orbital |

FIGURE 6.4 ▶ Contours for the $1s$, $2p_z$ and $3p_z$ orbitals.

3. As discussed in Section 6.1.3, we could replace ψ_{2p_z} with $-\psi_{2p_z}$, and this would be an equally valid stationary state (see Equation 6.9). So there is nothing wrong with writing the negative lobe on top and the positive on the bottom.

4. Different combinations of degenerate orbitals simplify different problems. For example, the normal representations p_x, p_y, p_z of the $l = 1$ orbitals are actually mixtures of different m_l values. The stationary states with a single value of $m_l \neq 0$ have real and imaginary parts, and are difficult to visualize.

6.4 MULTIELECTRON ATOMS AND MOLECULES

Schrödinger's equation is not analytically solvable for anything more complicated than a hydrogen atom. Even a helium atom, or the simplest possible molecule (H_2^+) requires a numerical calculation by computer. However, these calculations give results which agree extremely well with experiment, so their validity is not doubted.

Even though the hydrogen-atom wavefunctions are not exact solutions for multielectron atoms or molecules, they are often used for a good qualitative description. We will discuss here briefly the major features which change from the hydrogen problem. Every elementary textbook has dozens of full-color figures to illustrate molecular orbitals and hybrids, and we will not duplicate these figures here; we are aiming for an understanding of *why these results are reasonable*.

6.4.1 Ordering of Energy Levels

In a hydrogen atom, the orbital energy is determined exclusively by the principal quantum number n—all the different values of l and m_l are degenerate. In a multielectron atom, however, this degeneracy is partially broken: the energy increases as l increases for the same value of n.

We can illustrate this by comparing the energies of the $1s$, $2s$ and $2p$ orbitals for a helium atom, which has two electrons. The first electron goes into the $1s$ orbital. Thus the atom He^+ has an electronic probability distribution which is given by putting $Z = 2$ into Equation 6.24 above:

$$P(r) \text{ (for } He^+) = |\psi_{1s}(r)|^2 = C_1^2 \exp(-4r/a_0) \tag{6.27}$$

Recall from Section 3.2 that a spherical shell of charge has the same effect outside the shell as it would have if it were completely concentrated at its center; however, it has no effect (produces no net force) on a charge inside the shell. Thus, very far from the nucleus ($r \gg a_0$), the electron effectively neutralizes half of the $+2e$ charge of the helium nucleus. A second electron far from the nucleus would feel an effective net charge of $+e$. *Near* the nucleus, however ($r \ll a_0$), a second electron would be almost completely inside the $1s$ distribution, and would feel no net force in any direction from the first electron; thus it would feel an effective net charge of $+2e$ (Figure 6.5, left).

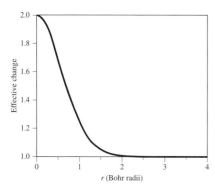
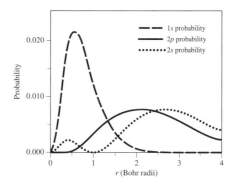

FIGURE 6.5 ▶ Left: effective charge in He^+ (nuclear charge minus the shielding from the one electron) as a function of distance from the nucleus. **Right**: probability of finding an electron at different positions for the $1s$, $2s$, and $2p$ orbitals of He^+.

Of course, a second electron will not be fixed at a single position. It will have a wavefunction with a spatial distribution, just as the first electron did. The right side of Figure 6.5 graphs the probability of finding an electron at different values of r for the $1s$, $2s$ and $2p$ orbitals. This probability is found by multiplying $|\psi|^2$ by the total volume between two shells, one with radius r, the other with radius $r + dr$. The $1s$ orbital is the lowest state, and a second electron goes into this orbital for the ground state of He (the two electrons have opposite values of m_s).

In hydrogen the $2s$ and $2p$ orbitals have the same energy. We can see qualitatively why this might be true: the $2s$ orbital has a small lobe very close to the nucleus, but the main lobe is farther from the nucleus than the bulk of the $2p$ orbital, and these effects exactly offset one another.

Even in a helium atom, however, the situation is different. The electron in the $1s$ orbital makes the effective charge greater for $r < a_0$, as shown in the left figure. This will have a more favorable effect on a $2s$ orbital than it will on a $2p$ orbital, because the $2s$ orbital is larger in that region. Hence the $2s$ orbital is lower in energy.

The same trend holds true for the relative ordering of the $3s$, $3p$, and $3d$ orbitals. In fact, in neutral atoms the $4s$ orbital is actually below the $3d$ orbital in energy; but the difference is generally small, and in transition metal ions the order is reversed. Thus the electronic configuration of calcium is [Ar] $4s^2$ (meaning it has the same first 18 electrons as argon, plus 2 $4s$ electrons), the configuration of Ti is [Ar] $4s^2 3d^2$, but the configuration of Ti^{2+}, isoelectronic with Ca, is instead [Ar] $3d^2$.

6.4.2 The Nature of the Covalent Bond

As we noted as the beginning of Chapter 5, one of the barriers to accepting Avogadro's postulate was the implication that molecules such as oxygen and hydrogen contained two identical atoms which were held together by some force. We are now in a position

FIGURE 6.6 ▶ **Left**: general arrangement of three bodies, in this case two protons and an electron. **Right**: minimum energy configuration for H_2^+. The electron is delocalized.

to understand why bonds between identical atoms are possible. We will concentrate on the simplest possible case: the molecule H_2^+ with two protons and one electron (Figure 6.6, left).

Two protons and an electron have a Coulombic potential energy given by:

$$U = \frac{1}{4\pi\varepsilon_0}\left(\frac{+e^2}{R} - \frac{e^2}{r_1} - \frac{e^2}{r_2}\right) \tag{6.28}$$

By inspection, the most favorable location for the electron is between the two protons. If the three particles lie on a line with the electron in the middle, then $r_1 = r_2 = R/2$:

$$U(R) = \frac{1}{4\pi\varepsilon_0}\left(\frac{+e^2}{R} - \frac{e^2}{(R/2)} - \frac{e^2}{(R/2)}\right) = \frac{1}{4\pi\varepsilon_0}\frac{-3e^2}{R} \tag{6.29}$$

Thus classically the potential energy gets more negative (more favorable) as the nuclear separation decreases. In fact, three charged balls would collapse together, and the "bond length" would be comparable to the diameter of the proton—about five orders of magnitude smaller than the experimentally measurable separation of 106 picometers.

By analogy with the hydrogen atom itself, you might have guessed what prevents this collapse—the Uncertainty Principle. As the distances shrink, the uncertainty in the electron's momentum (and hence its kinetic energy) must increase. For example, suppose we took the geometry on the right side of Figure 6.6 and cut all of the distances in half.

- From Equation 6.28, the new potential energy would be twice as negative.
- In order for the configuration to be stable, the electron must be between the protons. So Δx, Δy and Δz would all be cut in half as well.
- By the Uncertainty Principle, Δp_x, Δp_y, and Δp_z would all become twice as large.
- The overall length of the shortest possible momentum vector would double.
- The minimum possible kinetic energy $K = p^2/2m$ would quadruple.

Thus halving the size reduces the potential energy, but raises the kinetic energy. Eventually, as the internuclear separation gets shorter, the increase in kinetic energy has to more than offset the decrease in potential energy, and thus the overall energy will increase if the molecule shrinks further.

When the protons are separated by a large distance, the stationary states are the normal hydrogen orbitals centered on each proton, and the states on each atom have the same energy. Stationary states with the same energy (called **degenerate stationary states**) have a special feature in Schrödinger's equation. Suppose the wavefunctions $\psi_1(x)$ and $\psi_2(x)$ are two degenerate stationary states, both with energy E.

$$-\frac{\hbar^2}{2m}\frac{d^2\psi_1(x)}{dx^2} + U(x)\psi_1(x) = E\psi_1(x) \tag{6.30}$$

$$-\frac{\hbar^2}{2m}\frac{d^2\psi_2(x)}{dx^2} + U(x)\psi_2(x) = E\psi_2(x) \tag{6.31}$$

We can multiply Equation 6.30 by some coefficient c_1; multiply Equation 6.31 by some coefficient c_2; and add them together to get:

$$-\frac{\hbar^2}{2m}\frac{d^2(c_1\psi_1(x) + c_2\psi_2(x))}{dx^2} + U(x)\,(c_1\psi_1(x) + c_2\psi_2(x))$$
$$= E\,(c_1\psi_1(x) + c_2\psi_2(x)) \tag{6.32}$$

So $c_1\psi_1 + c_2\psi_2$ is also a solution to Schrödinger's equation with the same energy. The coefficients c_1 and c_2 have to be chosen to satisfy Equation 6.4, but otherwise they are arbitrary. So we can take any combination we want of stationary states with the *same* energy, and the combination is still a stationary state with that energy.

Combining atomic orbitals lets us create *molecular orbitals* which reinforce or weaken the probability in the energetically favored region between the protons by combining these atomic orbitals. For example, the two different $1s$ orbitals can be combined to create two new orbitals, called 1σ and $1\sigma^*$.

$$\psi_{1\sigma} = C_{1\sigma}(\psi_{1s,\,\text{atom 1}} + \psi_{1s,\,\text{atom 2}})$$
$$\psi_{1\sigma^*} = C_{1\sigma^*}(\psi_{1s,\,\text{atom 1}} - \psi_{1s,\,\text{atom 2}}) \tag{6.33}$$

The factors $C_{1\sigma}$ and $C_{1\sigma^*}$ are both about $\sqrt{2}$, and are needed to preserve the normalization. We must end up with the same number of combinations as the number of atomic orbitals we used. This can be understood by analogy with describing the distance between two particles in a plane by two different coordinate systems, rotated from one another by $45°$ (Figure 6.7).

If we pick any coordinate system and put one of the particles at the origin $(0, 0)$, the coordinates of the other particle (x, y) directly give the distance $r = \sqrt{x^2 + y^2}$. Rather than using a horizontal x-axis, we can create a new axis x' in any direction we want by combining x and y. But this new vector will only have the same length as did x and y if we divide by the appropriate factor. In the case shown here, where x and y are added with equal weight, the factor is $\sqrt{2}$. If we create a new x'-axis, we must also create a new y'-axis for the coordinates to give the right distance between the particles, and the y'-axis has to be constructed in such a way that it is perpendicular to x'.

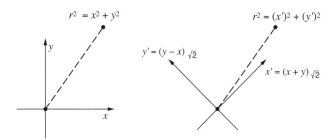

FIGURE 6.7 ▶ If we pick one point in a plane as the origin $(0, 0)$, and define two perpendicular directions as the x and y axes, the position of any other point can be specified by its coordinates (x, y). Any other choice of direction for the two axes (x' and y') will change the coordinates, but the distance will not change, so $x^2 + y^2 = x'^2 + y'^2$.

The orbital ψ_{1s} is positive everywhere, so the 1σ molecular orbital (called a ***bonding orbital***) is also positive everywhere, and has an enhanced probability of putting the electron between the two protons. This orbital is lower in energy than the $1s$ orbitals were by themselves. On the other hand, the $1\sigma^*$ orbital (called an ***antibonding orbital***) is higher in energy, because the probability of finding the electron in the energetically favored region is reduced. It has a *nodal plane* (a region where $\psi = 0$) which bisects the line between the two protons. As we saw with the particle-in-a-box wavefunctions, ***nodal surfaces*** are generally a good indication of increased energy. So as the separation between the two atoms decreases, the 1σ and $1\sigma^*$ states begin to be separated in energy, and in the H_2^+ molecule, the electron goes into the 1σ orbital. In the hydrogen molecule, a second electron goes into the same orbital, with the opposite value of m_s (just as in the helium atom). Additional electrons would have to go into the antibonding orbital, so the net energy gain from the bonds is reduced; in fact, the molecule He_2, which would need two electrons in 1σ and two in $1\sigma^*$ is not stable.

This description of the 1σ and $1\sigma^*$ orbitals is an oversimplification, since it is based on the $1s$ hydrogen-atom atomic orbitals, which are not stationary states for this more complicated problem. The shapes of the 1σ and $1\sigma^*$ orbitals (and the higher energy states) can be calculated by computer, and to some extent higher lying atomic orbitals have to be mixed in as well. However, the energy difference between the $1s$ state and the higher atomic orbitals is so large that the additions are quite small. For higher molecular orbitals, such as the ones used for bonding in molecules such as oxygen, the mixtures are more complicated; but the basic idea is still correct, and molecular orbitals are widely used to describe chemical bonding.

6.4.3 Hybridization

In a hydrogen atom, the $2s$ and $2p$ orbitals are degenerate. Therefore, as discussed in the last section, any other combination of these orbitals would also be a stationary state with the same energy. The only reason for writing these orbitals as we did in Equations 6.27 was mathematical convenience. In fact, as noted earlier, the orbital pictures in general chemistry books do *not* correspond to a well-defined value of m_l because those orbitals are complex functions. For example, the p_x orbitals are proportional to the sum of $m_l = +1$ and $m_l = -1$; p orbitals are proportional to the difference between $m_l = +1$ and $m_l = -1$ orbitals.

In forming molecules, it often makes sense to combine orbitals with different values of l as well, thus creating so-called **hybrid orbitals**. Consider, for example, the molecule BeH_2. Each hydrogen atom can contribute its $1s$ orbital to a molecular orbital, just as in the H_2^+ example in the last section. The beryllium atom in its ground state has two electrons in the $2s$ orbital, and since that orbital is already filled, it is not likely to contribute much stability to a molecular orbital (just as the filled $1s$ orbitals in two helium atoms did not create a bond for He_2).

The $2p$ orbitals still exist for a beryllium atom; they are simply empty. The atom could promote one electron to a $2p$ orbital, at the cost of some added energy. If this allowed it to contribute electron density to make two separate bonds, the added stability of the two bonds might be sufficient to justify the promotion energy.

The strengths of these bonds will be determined in large part by the overlap between the beryllium and hydrogen orbitals. Electrons repel one another, so multiple bonds will be most stable if the electron density is spread out over a wide region (without lengthening the bonds, which would decrease the overlap). A linear arrangement, say along the z-axis in space, with the beryllium in the center of the two hydrogens would achieve this. Unfortunately, the $2s$ orbital has equal electron density in all directions. Only the $2p_z$ orbital is aligned to overlap with either hydrogen, and the $2p_z$ orbital can only accept two of the four electrons available.

We can improve the overlap by combining the $2s$ and $2p_z$ orbitals to make two **sp-hybrid orbitals** (Figure 6.8):

$$\psi_{sp^+} = \frac{(\psi_{2s} + \psi_{2p_z})}{\sqrt{2}}$$

$$\psi_{sp^-} = \frac{(\psi_{2s} - \psi_{2p_z})}{\sqrt{2}} \qquad (6.34)$$

The $\sqrt{2}$ factor is for normalization, just as in Figure 6.3. One orbital points more to the right in Figure 6.8, so it can overlap well with the hydrogen atom to the right; the other orbital overlaps well with the hydrogen atom to the left. Notice that the *nodal surface* (the surface between the positive and negative lobes where $\psi = 0$) for these orbitals is neither a plane (as it is for the $2p$ orbitals) nor a sphere (as it is for the $2s$ orbital).

Strictly speaking, these sp hybrids are not stationary states, because they were created by combining orbitals with different energies. However, the energy difference is relatively small, and we can treat these hybrids as approximately valid. In fact, the two bonds in BeH_2 are identical and the molecule is linear, so the bonding orbital must look very much like what is pictured here.

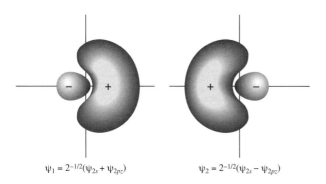

$$\psi_1 = 2^{-1/2}(\psi_{2s} + \psi_{2p_z})$$ $$\psi_2 = 2^{-1/2}(\psi_{2s} - \psi_{2p_z})$$

FIGURE 6.8 ▶ Two sp hybrid orbitals, formed by taking the sum or the difference of the $2s$ and $2p_z$ orbitals.

More complicated hybrid orbitals are used in many other problems. For example, the bonding in methane is best described by combinations of one s orbital and three p orbitals, which make a set of four equivalent orbitals which point to the corners of a tetrahedron:

$$\begin{aligned}
\psi_1 &= (\psi_{2s} + \psi_{2p_x} + \psi_{2p_y} + \psi_{2p_z})/2 \\
\psi_2 &= (\psi_{2s} - \psi_{2p_x} - \psi_{2p_y} + \psi_{2p_z})/2 \\
\psi_3 &= (\psi_{2s} - \psi_{2p_x} + \psi_{2p_y} - \psi_{2p_z})/2 \\
\psi_4 &= (\psi_{2s} + \psi_{2p_x} - \psi_{2p_y} - \psi_{2p_z})/2
\end{aligned} \qquad (6.35)$$

and in some cases, hybrids with d orbitals are used as well. These orbitals use the combination of s and p to make give the hybrid orbitals directionality, thus improving the overlap with orbitals from other atoms. They look similar to Figure 6.6, except that the decreased amount of s character in each orbital somewhat diminishes the imbalance between the positive and negative lobes.

Hybridization is a wonderful, intuitive concept, and is taught in every general chemistry class. It also has serious limitations, particularly for elements below the first full row of the periodic table. For example, the structure of H_2O (bond angle 104.5°) is generally explained using the sp^3 orbitals in Equation 6.35 as a starting point, and then assuming that extra repulsion between the two oxygen lone pairs reduces the bond angle. Oxygen in the ground state has the configuration $1s^2 2s^2 2p^2$, with $2s$ lower than $2p$ in energy; but creating bonds with a larger overlap with the hydrogen (making sp^3 orbitals instead of just using p orbitals) makes up for the energy lost by replacing a pair of nonbonding electrons in the $2s$ orbital with a nonbonding pair in a higher-energy sp^3 hybrid orbital.

By this argument, H_2S, H_2Se and H_2Te should have about the same structure as water. Instead, the bond angle in these three molecules is nearly 90°, as if the p orbitals were not hybridized at all! One important reason for the difference is the extra radial nodes in hybridized orbitals with $n > 2$. Even the $3p_z$ orbital has four lobes instead of two and the $3s$ orbital has two radial nodes, so it is much more difficult to create hybrids which produce a large, node-free region for a favorable σ bond. The energy gained by creating a slightly improved bonding region does not overcome the energy difference between $3s$ and $3p$.

 PROBLEMS ▶

6-1.⋆ Use Equation 6.1 to prove the following results:

(a) $\left|e^{i\theta}\right| = 1$

(b) $e^{i\theta}$ and $e^{-i\theta}$ are reciprocals of one another $(e^{i\theta} = 1/e^{-i\theta})$

(c) $e^{i\theta}$ and $e^{-i\theta}$ are complex conjugates of one another $(e^{i\theta} = (e^{-i\theta})^*)$

6-2. Use the Taylor series expansions in Chapter 2 to verify Equation 6.1.

6-3.⋆ Equation 6.11 can be applied to a wavefunction ψ which is already one of the stationary states, thus giving

$$\Psi(t) = e^{-iEt/\hbar}\Psi(0)$$

where E is the energy of the stationary state. Use this to show that the probability distribution of a stationary state is independent of time.

6-4. Use Equation 6.11 to show that changing the definition of the zero point of energy (which is arbitrary, because potential energy is included) by an amount Δ changes $\psi(t)$ by a factor $e^{-i\Delta t}$. Also show that this arbitrary choice has no effect on the probability distribution, or on the expectation values of position, momentum, or kinetic energy.

6-5.⋆ Use Equation 6.4 to verify that the expression for the particle-in-a-box wavefunction (Equation 6.20) is correctly normalized.

6-6. Explain why $\langle K \rangle \neq \langle p \rangle^2 /2m$ for the particle in a box.

6-7.⋆ Graph the probability $P(x)$ for wavefunction $\Psi = \frac{1}{\sqrt{2}}(\psi_1 - \psi_2)$, where ψ_1 and ψ_2 are the first and second stationary states for the particle in a box (Equation 6.20).

6-8. Graph the wavefunction $\Psi = \frac{1}{\sqrt{2}}(\psi_1 + i\psi_2)$, where ψ_1 and ψ_2 are the first and second stationary states for the particle in a box (Equation 6.20).

6-9.⋆ Graph the wavefunction $\Psi = \frac{1}{\sqrt{2}}(\psi_1 + \psi_3)$, where ψ_1 and ψ_3 are the first and third stationary states for the particle in a box (Equation 6.20), and verify that it satisfies Equation 6.4 (the definite integral needed to verify this can be found in Appendix B). Without doing any explicit integrals, determine $\langle x \rangle$ and $\langle p \rangle$.

6-10. Equation 6.11 can be used to solve for the time evolution of a superposition state. Suppose we start at time $t = 0$ with the superposition state in Figure 6.3, $\Psi(t = 0) = \frac{1}{\sqrt{2}}(\psi_1 - \psi_2)$. We can simplify the mathematics a bit by choosing the energy of the bottom of the well as $U = -h^2/8mL^2$ instead of zero; this makes the total energy of the lowest state (kinetic plus potential) equal to zero ($E_1 = 0$) and makes $E_2 = 3h^2/8mL^2$.

(a) Write out an explicit expression for $\Psi(t)$ which is valid at all later times t, and an expression for the probability distribution $P(t)$. Verify that $P(t)$ is real and nonnegative everywhere.

(b) Use Equation 6.5 and Equation B.25 in Appendix B to evaluate the expectation value of the position, $\langle x \rangle$, at any time t.

(c) Use Equation 6.6 and Equation B.26 in Appendix B to evaluate the expectation value of the momentum, $\langle p \rangle$, at any time t.

(d) Your results in parts (b) and (c) above should show that $\langle p \rangle$ vanishes when $\langle x \rangle$ is at a maximum or minimum, and $\langle x \rangle = L/2$ when $\langle p \rangle$ is at a maximum or minimum. Explain this classically.

6-11. (a) The harmonic oscillator (mass m, potential energy $U = kx^2/2$) is also an exactly solvable problem in quantum mechanics. Substitute this form for U into Schrödinger's equation (Equation 6.8) to show that $\psi = Ce^{-x^2\sqrt{mk}/2\hbar}$ is a stationary state (C is just the normalization constant).

(b)* Find the energy of this stationary state. If you have the correct expression, it will look simpler if you introduce $\omega_0 = \sqrt{k/m}$, where ω_0 is the classical vibrational frequency discussed in Chapter 3.

(c) The stationary state ψ is actually the lowest energy solution for the harmonic oscillator, and can be applied to diatomic molecules (substituting the reduced mass μ for the mass m, as in Chapter 3). This lowers the dissociation energy from what would be predicted classically, because a molecule cannot be sitting in the bottom of the potential well. For the hydrogen molecule, use the force constant and well depth from Table 3.2 to verify the actual minimum dissociation energy.

6-12. The probability of finding an electron between r and $r + dr$ from the nucleus (graphed in Figure 6.3 for the He^+ atom) is given by squaring the wavefunction and multiplying by r^2. The most probable value is found by taking the derivative of this expression.

Calculate the most probable value of r for a $1s$ electron, a $2s$ electron and a $2p$ electron.

6-13.* Here is yet another bizarre result of quantum mechanics for you to ponder. The $1s$ wavefunction for a hydrogen atom is unequal to zero at the origin. This means that there is a small, but nonzero probability that the electron is inside the proton. Calculation of this probability leads to the so-called "hyperfine splitting"—the magnetic dipoles on the proton and electron interact. This splitting is experimentally measurable. Transitions between the hyperfine levels in the $1s$ state of hydrogen are induced by radiation at 1420.406 MHz. Since this frequency is determined by nature, not man, it is the most common choice as a transmission frequency or a monitored frequency in searches for extraterrestrial intelligence.

The proton has a diameter of approximately 10^{-15} m. Explain why it is not possible to design an experiment which would measure the location of a $1s$ electron and find it to be inside the nucleus.

6-14. Use the hydrogen wavefunctions to find the value of z where the wavefunction $\psi_{sp^+} = (\psi_{2s} + \psi_{2p_z})/\sqrt{2}$ is most positive, and the position where it is most negative.

Chapter 7

The Kinetic Theory of Gases

Nothing exists except atoms and empty space; everything else is opinion.

Democritos (ca. 460–370 BC)

Every chemistry student is familiar with the "ideal gas equation" $PV = nRT$. It turns out that this equation is a logical consequence of some basic assumptions about the nature of gases. These simple assumptions are the basis of the *kinetic theory of gases*, which shows that the collisions of individual molecules against the walls of a container creates pressure. This theory has been spectacularly successful in predicting the macroscopic properties of gases, yet it really uses little more than Newton's laws and the statistical properties discussed in the preceding chapters.

The ideal gas law has been used in many examples in earlier chapters, and some of the important physical properties of gases (the one-dimensional velocity distribution, average speed, and diffusion) were presented in Chapter 4. This chapter puts all of these results into a more comprehensive framework. For example, in Section 7.3 we work out how the diffusion constant scales with pressure and temperature, and we explore corrections to the ideal gas law.

7.1 COLLISIONAL DYNAMICS

Collisions provide a particularly simple application of Newton's laws, which we discussed in Chapter 3, because we can often say that long before and long after the collision (when the colliding bodies are separated by large distances) that the forces between them are zero. It is also often a good approximation to say that the forces during the collision did not significantly reduce the kinetic energy (such collisions are called *elastic*). Then for a collision between two particles with masses m_1 and m_2 we have *conservation of momentum*:

$$\vec{p}_{tot} = m_1 \vec{v}_{1,\text{initial}} + m_2 \vec{v}_{2,\text{initial}} = m_1 \vec{v}_{1,\text{final}} + m_2 \vec{v}_{2,\text{final}} \tag{7.1}$$

which is actually three separate equations in x, y, and z:

$$
\begin{aligned}
p_{x,\text{tot}} &= m_1 v_{x1,\text{initial}} + m_2 v_{x2,\text{initial}} = m_1 v_{x1,\text{final}} + m_2 v_{x2,\text{final}} \\
p_{y,\text{tot}} &= m_1 v_{y1,\text{initial}} + m_2 v_{y2,\text{initial}} = m_1 v_{y1,\text{final}} + m_2 v_{y2,\text{final}} \\
p_{z,\text{tot}} &= m_1 v_{z1,\text{initial}} + m_2 v_{z2,\text{initial}} = m_1 v_{z1,\text{final}} + m_2 v_{z2,\text{final}}
\end{aligned}
$$

We also have *conservation of energy*:

$$
\begin{aligned}
E &= \left(m_1 \left| \vec{v}_{1,\text{initial}} \right|^2 + m_2 \left| \vec{v}_{2,\text{initial}} \right|^2 \right) / 2 \\
  &= \left(m_1 \left| \vec{v}_{1,\text{final}} \right|^2 + m_2 \left| \vec{v}_{2,\text{final}} \right|^2 \right) / 2
\end{aligned}
\tag{7.2}
$$

which can be written in Cartesian coordinates as

$$
\begin{aligned}
E &= m_1 \left(v_{x1,\text{initial}}^2 + v_{y1,\text{initial}}^2 + v_{z1,\text{initial}}^2 \right) / 2 \\
  &\quad + m_2 \left(v_{x2,\text{initial}}^2 + v_{y2,\text{initial}}^2 + v_{z2,\text{initial}}^2 \right) / 2 \\
  &= m_1 \left(v_{x1,\text{final}}^2 + v_{y1,\text{final}}^2 + v_{z1,\text{final}}^2 \right) / 2 \\
  &\quad + m_2 \left(v_{x2,\text{final}}^2 + v_{y2,\text{final}}^2 + v_{z2,\text{final}}^2 \right) / 2
\end{aligned}
$$

Equations 7.1 and 7.2 do not assume any particular form for the potential energy, except that Equation 7.2 only applies at times when the interaction energy is negligible (long before or after the collision). In most cases, they do not completely determine the final trajectories, and the interaction potential has to be included to get a complete answer.

The simplest models view the interacting bodies as "hard spheres" (e.g., billiard balls). Mathematically, if r is the separation between the center of two molecules, we write the potential energy of interaction between them as:

$$U(r) = \infty \; (0 \le r \le \sigma); \; U(r) = 0 \; (r > \sigma) \tag{7.3}$$

In this case σ is the distance of closest approach between the centers of the two molecules, which is the same as the diameter of a single molecule. This potential generates no forces for $r > \sigma$. The discontinuity at $r = \sigma$ implies an infinitely large force (and hence collisions are instantaneous). There are really no perfect hard spheres, but this approximation often simplifies calculations dramatically, and often gives good approximate results.

It is worth illustrating the uses of Equations 7.1 and 7.2 with a few examples. Suppose two balls, each with mass 1 kg and diameter $\sigma = 0.1$ m, are moving along the

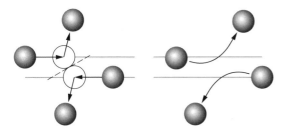

FIGURE 7.1 ▶ Collisions between two particles that interact through a hard-sphere potential (left) can be described by conservation of momentum and energy. A potential that is more realistic for atoms and molecules (right) changes the trajectories.

x-axis with equal and opposite speeds of 10 m · s^{-1} (Figure 7.1, left). After the balls collide (elastically), what will be their speeds and directions?

The total momentum and kinetic energy are:

$$
\begin{aligned}
\left|\vec{p}_{\text{tot,initial}}\right| &= \left|\vec{p}_{1,\text{initial}} + \vec{p}_{2,\text{initial}}\right| \\
&= (1 \text{ kg})(+10 \text{ m} \cdot \text{s}^{-1}) + (1 \text{ kg})(-10 \text{ m} \cdot \text{s}^{-1}) \\
&= 0 \text{ kg} \cdot \text{m} \cdot \text{s}^{-1} = \left|\vec{p}_{\text{tot,final}}\right| \quad\quad (7.4)\\
E_{\text{initial}} &= \frac{(1 \text{ kg})(+10 \text{ m} \cdot \text{s}^{-1})^2}{2} + \frac{(1 \text{ kg})(-10 \text{ m} \cdot \text{s}^{-1})^2}{2} \\
&= 100 \text{ J} = E_{\text{final}}
\end{aligned}
$$

Since the masses are equal, the only way $\vec{p}_{\text{tot,final}}$ can vanish is if $\vec{v}_{1,\text{final}} = -\vec{v}_{2,\text{final}}$. Since the speeds are then equal, Equation 7.4 shows that both *speeds* must be 10 m · s^{-1}, but the direction is unknown. For hard spheres the additional information needed to determine the direction comes from the ***impact parameter b***, defined as the minimum distance of separation between the centers of the two balls if they were to follow their initial trajectories. In Figure 2.4, the impact parameter is the distance between the two dotted lines.

If $b = 0$ the balls hit head on, and reverse direction. If $b > 0$ the balls do not interact at all. For intermediate values of b the balls are deflected into different directions; any direction is possible.

The trajectories are still more complicated if the balls interact with a realistic potential. The right side of Figure 7.1 illustrates a case that is realistic for atoms and molecules, as we discussed in Chapter 3: the interaction potential is attractive at long distances and repulsive at short distances.

Now suppose the two colliding partners have quite different masses. For example, suppose a helium atom (mass 6.64×10^{-27} kg) is traveling perpendicular to the flat wall of a container (mass 1 kg) at a speed of 1000 m · s^{-1}. Choose the z-axis as the initial direction of motion of the helium atom. The motion after the collision will also be along the z-axis, since the atom is moving perpendicular to the wall—no force is ever exerted in the x or y directions. Now the momentum and velocity vectors have only one nonzero component (the z component), so we can write the momentum and velocity as numbers instead of vectors.

The momentum and kinetic energy conservation equations are then:

$$
\begin{aligned}
p_{\text{tot,initial}} &= (6.64 \times 10^{-27} \text{ kg})(1000 \text{ m} \cdot \text{s}^{-1}) \\
&= 6.64 \times 10^{-24} \text{ kg} \cdot \text{m} \cdot \text{s}^{-1} \\
&= p_{\text{He,final}} + p_{\text{wall,final}} \\
&= (6.64 \times 10^{-27} \text{ kg}) v_{\text{He,final}} + (1 \text{ kg}) v_{\text{wall,final}}
\end{aligned}
\tag{7.5}
$$

$$
\begin{aligned}
E &= \frac{(6.64 \times 10^{-27} \text{ kg})(1000 \text{ m} \cdot \text{s}^{-1})^2}{2} = 3.3 \times 10^{-21} \text{ J} \\
&= \frac{(6.64 \times 10^{-27} \text{ kg})(v_{\text{He,final}})^2}{2} + \frac{(1 \text{ kg})(v_{\text{wall,final}})^2}{2}
\end{aligned}
\tag{7.6}
$$

We can do an exact algebraic solution to Equations 7.5 and 7.6 (Problem 7-1), or just look at these equations and simplify things dramatically. Because of the large difference in mass, it is obvious that the wall does not move very fast after the collision. In fact, if *all* of the kinetic energy went into the wall, its speed would be $s = \sqrt{2E/m} = 8.1 \times 10^{-11}$ m $\cdot$ s^{-1}. But if that were the case, the final momentum would be 8.1×10^{-11} kg $\cdot$ m $\cdot$ s^{-1}—almost thirteen orders of magnitude too large. The only way to conserve the total momentum is to make almost all of the kinetic energy stay in the helium atom, which thus bounces back with a speed very close to 1000 m $\cdot$ s^{-1} (Figure 7.2, top).

Since the helium atom's direction has reversed, $p_{\text{He,final}} = -6.64 \times 10^{-24}$ kg $\cdot$ m $\cdot$ s^{-1}. Conservation of momentum implies that $p_{\text{wall,final}} = 1.33 \times 10^{-23}$ kg $\cdot$ m $\cdot$ s^{-1} to balance Equation 7.5. The wall acquires a kinetic energy of $p_{\text{wall,final}}^2/2m = 8.8 \times$

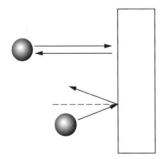

FIGURE 7.2 ▶ When a light particle collides with a heavy wall, very little energy is transferred. The component of the velocity perpendicular to the wall is reversed; the other components are unaffected.

10^{-47} J. We can now be more quantitative—all but about 10^{-23} of the kinetic energy stays in the atom.

Notice that we did not assume a specific functional form for the interaction potential, except to assume that the direction of the force exerted by the wall is perpendicular to the wall, hence opposite to the initial velocity vector. If the helium atom is approaching from some other direction, the component of the atom's velocity perpendicular to the wall will be reversed but the other components are unaffected (as shown at the bottom of Figure 7.2).

7.2 PROPERTIES OF IDEAL GASES

7.2.1 Assumptions Behind the Ideal Gas Law

The simplest treatment of the properties of gases starts with the following *assumptions*, which will determine the limits of validity of the ideal gas equation $PV = nRT$:

1. Gases are mostly empty space at normal pressures. At standard temperature and pressure (STP; 1 atm pressure, temperature 273K), the same amount of matter will occupy about 1000 times more volume if it is in the gaseous state than if it is a solid or liquid. At much higher densities (for example, pressures of several hundred atmospheres at 273K) this assumption will not be valid.

2. Intermolecular forces between gas molecules are assumed to be negligible, and collisions between gas molecules are ignored (in our initial treatment). Collisions with the container are assumed to be elastic, meaning that both momentum and energy are conserved, as described in Section 7.1 above.

3. Gas molecules are constantly moving, with a random distribution of directions and speeds. This is also a very reasonable assumption, unless the molecules in the gas were prepared in some way which (for example) made all of them move at nearly the same speed—and even then, collisions with the walls would randomize the distribution in practice. This distribution of speeds will be found using the Boltzmann distribution we discussed in Chapter 4.

4. The gas has evolved to a state, commonly called *equilibrium*, where none of the macroscopic observables of the system (temperature, pressure, total energy) are changing. Equilibrium is a *macroscopic* and *statistical* concept, as noted in Chapter 4. For example, the pressure cannot be perfectly constant, since this pressure comes from molecules hitting the wall of the container. If we look on a fine enough time scale (for example, 1 femtosecond at a time) the number of collisions cannot be exactly constant. However, as we will see, the number of collisions per second against the wall is so large that replacing the instantaneous (rapidly fluctuating) pressure with its average is a reasonable assumption.

7.2.2 Calculating Pressure

Pressure is defined as force per unit area. The pressure exerted by a gas comes from the forces exerted by collisions of gas molecules with the walls of the container. Since the mass of the walls of the container is much larger than the mass of each particle, the assumption of elastic collisions implies that the velocity component perpendicular to the wall is exactly reversed, and the other two components are unaffected as discussed in Section 7.1.

Let us consider what this means for one molecule, which we will initially assume is moving at $t = 0$ in the $+y$ direction ($v_x = v_z = 0$) in a cubic box of side length L (Figure 7.3). The molecule hits the right wall with velocity v_y and bounces back with velocity $-v_y$. The y-component of the momentum, p_y, is the product of mass times v_y, so we have

$$p_{\text{initial}} = +mv_y \quad p_{\text{final}} = -mv_y \quad (\Delta p)_{\text{molecule}} = -2mv_y \qquad (7.7)$$

Since total momentum is conserved, this implies the wall also picks up momentum to compensate for the molecules' change in momentum,

$$(\Delta p)_{\text{wall}} = -(\Delta p)_{\text{molecule}} = 2mv_y \qquad (7.8)$$

This change in the wall's momentum in a very short time (the duration of the collision Δt_{coll}) implies the wall exerts a large force ($F = \Delta p / \Delta t_{\text{coll}}$). If it is not obvious to you that $F = \Delta p / \Delta t$, review Section 3.1; $F = dp/dt$ is actually Newton's Second Law. We use "Δp" and "Δt" instead of the differentials dp and dt because the collisions, while brief, are not infinitesimal. If gas molecules and container walls really were incompressible, they would be in contact with the wall for an infinitesimal time, and the force would have to be infinite.

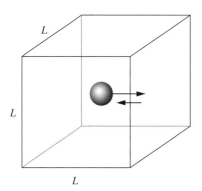

FIGURE 7.3 ▶ Simple model system for derivation of the ideal gas law. We start by assuming there is only one molecule in a cube of length L, and it is moving directly along the y-axis. We will generalize these results in the next section.

In reality, atoms and molecules (and container walls, made of atoms and molecules) can be somewhat compressed, and collisions last for a time on the order of picoseconds. The force is concentrated on a spot which is about the same size as the molecule (typically on the order of 1 nm^2) which still makes the pressure (force per unit area) at the point of impact quite high (Problem 7-2). So the wall deforms briefly, but in general the impact is not large enough to break chemical bonds, so it quickly recovers its shape.

Now we will switch over to a macroscopic view, looking at the average force and pressure, rather than the instantaneous values. Collisions with the right wall (and thus force on the wall) happen every time the molecule makes a round trip between the left and right walls, thus traveling a total distance $2L$; the time between collisions is then $2L/v_y$. So the *average force* on the right wall from these collisions is the momentum transferred to the wall per unit time:

$$F = \frac{2mv_y}{2L/v_y} = \frac{mv_y^2}{L} \tag{7.9}$$

The force per unit area (pressure) can now be written as

$$P = \frac{F}{A} = \frac{mv_y^2}{L^3} = \frac{mv_y^2}{V} \tag{7.10}$$

since $A = L^2$, and $L \times A$ (length times area) is the volume of the box V.

If the molecules do not interact with each other, the pressure exerted by the ith molecule is independent of the pressure exerted by any other molecule, and the total pressure is the sum of the pressures exerted by each molecule. Thus:

$$P = \frac{m}{V} \sum_{i=1}^{N} v_y^2 = \frac{Nm}{V} \overline{v_y^2} \tag{7.11}$$

Of course, not all of the molecules are moving straight along the y-axis. If v_x or $v_z \neq 0$, molecules also bounce off the top, bottom, front, or back walls (Figure 7.4). But those collisions only change v_z (if the molecule hits the top or bottom) or v_x (if the molecule hits the front or back). The rate of collisions with the left wall and the momentum transfer to that wall are unaffected. So Equation 7.11 is independent of the initial velocity directions of the individual molecules. It is also independent of the size or shape, although we would have to make a more complicated argument for some other geometry.

In general, we expect the velocity distributions in the x-, y-, and z-directions to be the same, and thus we can relate the pressure to the mean-squared speed $\overline{s^2}$:

$$\overline{v_x^2} = \overline{v_y^2} = \overline{v_z^2}; \quad \overline{s^2} = \overline{v_x^2} + \overline{v_y^2} + \overline{v_z^2} = 3\overline{v_y^2} \tag{7.12}$$

The total energy is given by:

$$E = \sum_{i=1}^{N} \frac{ms_i^2}{2} = N \left(\frac{m\overline{s^2}}{2} \right) = \frac{3Nm\overline{v_y^2}}{2} \tag{7.13}$$

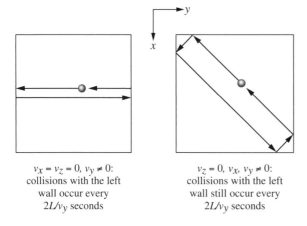

$v_x = v_z = 0, v_y \neq 0$:
collisions with the left
wall occur every
$2L/v_y$ seconds

$v_z = 0, v_x, v_y \neq 0$:
collisions with the left
wall still occur every
$2L/v_y$ seconds

FIGURE 7.4 ▶ Only the velocity component perpendicular to a wall contributes to the pressure, since other components do not change the round-trip time.

so we have:

$$PV = \frac{2}{3}E \tag{7.14}$$

The product of pressure and volume has units of energy—which normally is not obvious because we usually think of volume in liters and pressure in pascals or atmospheres. But this will become very important later in considering energy and work.

7.2.3 The One-Dimensional Velocity Distribution and the Ideal Gas Law

In order to get from Equation 7.14 to the ideal gas law, we need to relate v_y^2 to the temperature. As discussed in Chapter 4, we can use the Boltzmann distribution, Equation 4.26, to give the probabilities of observing different velocities:

$$P(v_y)\,dv_y = \left\{\sqrt{m/2\pi kT}\right\} e^{-mv_y^2/(2k_BT)}\,dv_y \tag{7.15}$$

The term in {brackets} in Equation 7.15 is the normalization constant, chosen so that $\int P(v)\,dv = 1$. Equation 7.15 is a Gaussian peaked at $v_y = 0$ with standard deviation $\sigma = \sqrt{kT/m}$. We also showed in Chapter 4 that

$$\overline{v_y^2} = \frac{k_BT}{m}; \quad \left(\overline{v_y^2}\right)^{1/2} = \sqrt{\frac{k_BT}{m}} \tag{7.16}$$

$$E = \frac{3}{2}Nk_BT = \frac{3}{2}NRT \tag{7.17}$$

where n is the number of moles, and the relation between R and k_B is $R = N_{\text{Avogadro}}k_B = (6.022 \times 10^{23})k_B = 8.3144$ J/(mole · K). Combining Equations 7.14 and 7.17 gives

$$PV = Nk_BT = nRT \tag{7.18}$$

7.2.4 The Three-Dimensional Speed Distribution

So far we have only talked about the velocity in a single direction, but Equation 7.12 also lets us relate this velocity to the speed:

$$\overline{s^2} = \frac{3k_BT}{m}; \quad \left(\overline{s^2}\right)^{1/2} = \sqrt{\frac{3k_BT}{m}} \tag{7.19}$$

Equation 7.19 gives a root-mean-squared speed ("rms speed") at 273K of 1840 m/s for H_2, 493 m/s for N_2, and 206 m/s for Br_2. To get the rms component in any particular direction (for example, $\left(\overline{v_y^2}\right)^{1/2}$, divide by $\sqrt{3}$.

The properties of the three-dimensional *speed* distribution can be readily derived as well. This speed distribution turns out to be:

$$N(s) \propto s^2 \exp(-ms^2/2k_BT) \tag{7.20}$$

which looks very similar to Equation 7.15 except for an extra factor of s^2. This extra factor comes because the number of different ways to get a particular speed goes up as we increase the speed. To see this, let us draw a "velocity space" where the coordinates correspond to the velocity components along x-, y-, or z-axes. Thus every possible velocity corresponds to a single point (v_x, v_y, v_z) in this three-dimensional space. In this velocity space, the *surface of a sphere* corresponds to all molecules with the same speed, but traveling in different directions (Figure 7.5). Thus there is a purely geometric factor—the fact that shell volume increases as $|\vec{v}|$ increases—that adds an additional term to the velocity distribution.

Another way to see this is to assume, for the moment, that the velocity in the x-, y-, or z-directions is restricted to discrete values (say integral numbers of meters per second). Then the number of ways to have a speed between 10 m · s^{-1} and 11 m · s^{-1} is the number of solutions to the equation

$$10^2 < v_x^2 + v_y^2 + v_z^2 < 11^2 \quad (x, y, z \text{ integers}) \tag{7.21}$$

There are many solutions to this equation. For example, just for $v_x = \pm 10$ m · s^{-1}, we can have these values of (v_x, v_y, v_z) and many others:

$(\pm 10, 0, 0)$	$(\pm 10, \pm 1, 0)$	$(\pm 10, 0, \pm 1)$	$(\pm 10, \pm 1, \pm 1)$	$(\pm 10, \pm 2, 0)$	$(\pm 10, 0, \pm 2)$
$(\pm 10, \pm 1, \pm 2)$	$(\pm 10, \pm 2, \pm 1)$	$(\pm 10, \pm 2, \pm 2)$	$(\pm 10, \pm 3, 0)$	$(\pm 10, 0, \pm 3)$	$(\pm 10, \pm 3, \pm 1)$
$(\pm 10, \pm 1, \pm 3)$	$(\pm 10, \pm 2, \pm 3)$	$(\pm 10, \pm 3, \pm 2)$	$(\pm 10, \pm 3, \pm 3)$	$(\pm 10, \pm 4, 0)$	
$(\pm 10, 0, \pm 4)$	$(\pm 10, \pm 4, \pm 1)$	$(\pm 10, \pm 1, \pm 4)$	$(\pm 10, \pm 4, \pm 2)$	$(\pm 10, \pm 2, \pm 4)$	

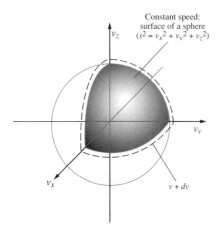

FIGURE 7.5 ▶ Schematic illustration of a "velocity space." Each possible velocity corresponds to a single point (v_x, v_y, v_z) in three dimensions. The total number of possible velocities (the volume) in a shell near some speed s is proportional to s^2. Thus the three-dimensional speed distribution has an extra factor of s^2.

There are a total of 1328 solutions. However, there are only 380 solutions for a speed between $5 \text{ m} \cdot \text{s}^{-1}$ and $6 \text{ m} \cdot \text{s}^{-1}$; there are 5240 solutions for a speed between $20 \text{ m} \cdot \text{s}^{-1}$ and $21 \text{ m} \cdot \text{s}^{-1}$. The number of solutions scales approximately as s; with a finer grid of discrete values (say tenths or hundredths of a meter per second) the s scaling would be nearly exact.

$P(s)\,ds$ gives the probability of finding the speed to be between s and $s+ds$. Therefore it must be normalized, $\int P(s)\,ds = 1$. The integral is given in Appendix B, and can be used to show that:

$$P(s)\,ds = 4\pi \left\{ \frac{m}{2\pi k_B T} \right\}^{3/2} s^2 (e^{-ms^2/2k_B T})\,ds \tag{7.22}$$

Note that the probability of finding $v \approx 0$ is very low because of the geometric factor. The most probable speed $s_{\text{most probable}}$ is found by differentiation of $P(s)$

$$\left. \frac{dP(s)}{ds} \right|_{s=s_{\text{most probable}}} = 0$$

which gives (Problem 7-3):

$$s_{\text{most probable}} = \sqrt{2k_B T/m} \tag{7.23}$$

Note that $s_{\text{most probable}}$ is not exactly the same as $\left(\overline{s^2}\right)^{1/2}$, which we found to be $\sqrt{3k_B T/m}$ (Equation 7.19). For that matter, we could find the average speed $\bar{s}$ to be $\sqrt{8k_B T/\pi m}$ using Equation 4.15. The differences between these measures of the distribution are small, but real.

7.2.5 Other ideal Gas Properties

▶ *Mixture Velocities and Effusion*

In a mixture of two or more gases, collisions eventually make the molecules achieve the same temperature, and hence the same kinetic energy. However, this implies that they have different velocities. If the two gases are labeled A and B, Equation 7.19 implies

$$\frac{\left(\overline{s_A^2}\right)^{1/2}}{\left(\overline{s_B^2}\right)^{1/2}} = \sqrt{\frac{m_B}{m_A}} \tag{7.24}$$

The rate at which molecules "pour out" of a small hole into a vacuum (see Figure 7.6) is proportional to the velocity of the molecules in the direction of the hole (or, equivalently, if the hole were filled with a plug, it is proportional to the rate at which molecules hit the plug). This process, which creates a "molecular beam" into the vacuum, is called *effusion*. A mixture of two gases with two different masses will have different mean velocities in the direction of the wall, and the lighter gas will leave the box more rapidly.

Gaseous effusion can be used to separate different masses. For example, the Manhattan project during World War II required separation of uranium isotopes, since only about 0.7% of naturally occurring uranium is the fissionable isotope ^{235}U. It was separated from the 99.3% abundant isotope ^{238}U by making the two hexafluorides and exploiting effusion through porous membranes. The mass difference is extremely small, so the effusing gas is only very slightly enriched in ^{235}UF$_6$; the process has to be repeated many times to produce a large enrichment.

Molecular beams are very important tools for characterizing intermolecular and intramolecular reactions. In fact, the 1988 Nobel Prize in Chemistry was awarded to Yuan Lee, Dudley Herschbach, and John Polanyi for studies which were mostly made possible by this technique. A particularly useful variant is the *supersonic molecular beam*, which in the simplest case pushes a high-pressure mixture of helium and trace amounts of some larger "guest" molecule through a nozzle. When the helium atoms enter the

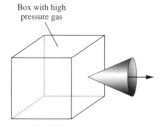

Box with high
pressure gas

FIGURE 7.6 ▶ Gases "pour out" of a box into an evacuated region at rates proportional to their speeds, hence proportional to $1/\sqrt{m}$. Thus a mixture of gases with different masses can be separated by this method.

vacuum to form the supersonic beam, they convert most of their random energy into kinetic energy in one direction. This also cools the guest molecules, so reactions can be monitored without complications from the wide range of initial energies associated with chemical reactions in a bulb or a flame.

▶ Heat Capacity

Equation 7.17 shows that the total kinetic energy of one mole of a monatomic gas is $E = 3RT/2$. Unfortunately the total energy of a system is a difficult quantity to measure directly. It is much easier to measure heat capacities—for example, the number of joules necessary to raise the temperature of one mole of gas by one degree Kelvin.

Because of the ideal gas relation $PV = nRT$, we cannot raise the temperature while keeping both the pressure and volume constant. Chemical reactions are commonly done under one of two limiting conditions: constant volume (for example, in a steel container) or constant pressure (for example, in a balloon or in a piston which can expand). The heat capacity depends on what is kept constant. Expanding a gas changes the potential energy. Raising the temperature of gas inside a piston will push the piston out, thus doing work on the surroundings, so only a portion of the added heat will be converted into a temperature increase.

Equation 3.12, the work-energy theorem, can be converted into a more convenient form for a gas in a piston in the geometry, as we showed in Equation 3.14:

$$U(r_2) - U(r_1) = - \int_{r=r_1}^{r=r_2} \left(-P_{ext} \sum A \right) dr$$

$$= \int P_{ext} \sum (A \, dr) = \int_{V=V_1}^{V_2} P_{ext} \, dV$$

If the volume is held constant, the integral vanishes. The added energy required to heat one mole of gas from temperature T_1 to temperature T_2 ($q(T_2) - q(T_1)$), and the *constant-volume molar heat capacity* c_v, are given by:

$$q(T_2) - q(T_1) = E(T_2) - E(T_1) = \frac{3R}{2}(T_2 - T_1);$$

$$c_V = \left(\frac{dq}{dT} \right)_V = \frac{3R}{2} \tag{7.25}$$

The subscripted V is there to remind us that the volume is assumed to remain constant. If the pressure is instead held constant (at the external pressure), the work done from

Equation 3.14 is $P(V_2 - V_1)$, so the added energy becomes:

$$\begin{aligned} q(T_2) - q(T_1) &= E(T_2) - E(T_1) + \{P(V_2 - V_1)\} \\ &= \frac{3R}{2}(T_2) - \frac{3R}{2}(T_1) + \{R(T_2 - T_1)\} \qquad (7.26) \\ &= \frac{5R}{2}(T_2 - T_1) \end{aligned}$$

where the term in brackets is converted from pressure and volume to temperature using the ideal gas law. Sometimes Equation 7.26 is written in a different form by defining the **enthalpy** $H = E + PV$

$$\begin{aligned} q(T_2) - q(T_1) &= E(T_2) - E(T_1) + \{P(V_2 - V_1)\} \\ &= H(T_2) - H(T_1) = \frac{5R}{2}(T_2 - T_1) \qquad (7.27) \end{aligned}$$

Thus the **constant-pressure heat capacity** c_p is given by

$$c_p = \left(\frac{dq}{dT}\right)_q = \left(\frac{dH}{dT}\right)_p = c_v + R \qquad (7.28)$$

Equations 7.25 and 7.28 agree very well with experiments for monatomic gases. The relation between c_v and c_p in Equation 7.28 also works for polyatomic gases, but calculating c_v and c_p requires a much more sophisticated treatment which explicitly includes the vibrational energy levels. As noted in Chapter 5, for virtually all diatomic gases at room temperature $c_v \approx 5R/2$.

▶ Speed of Sound

Another quantity which is closely related to the average molecular speed is the **speed of sound**, which we will write as s_{sound}. As noted in Chapter 3, sound waves are actually waves of gas pressure (Figure 3.6)—the density of gas molecules is alternately slightly higher or lower than the equilibrium value. This disturbance travels at a characteristic speed which is clearly not very different from the average molecular speed, but getting the precise numerical value requires a fairly sophisticated treatment. The answer turns out to be

$$s_{sound} = \sqrt{\frac{\gamma k_B T}{m}} \qquad (7.29)$$

where $\gamma = c_p/c_v$. From the discussion in the last section, $\gamma = 5/3$ for a monatomic gas and $\gamma \approx 1.41$ for a diatomic gas. The quantity γ also appears in some other equations for ideal gas properties, such as in the expression for **adiabatic expansion** (see Problem 7-11).

In air, s_{sound} is approximately 330 m/sec. A common lecture trick involves speaking into a bag filled with a helium/oxygen mixture, which is breathable but raises the speed of sound dramatically compared to air (essentially a nitrogen/oxygen mixture):

$$\frac{s_{sound}(He)}{s_{sound}(N_2)} \approx 2.9 \qquad (7.30)$$

Since the speed rises while your "voice box" remains the same size, the frequency $\nu = s_{sound}/\lambda$ of the sound disturbances you make goes up. As the sound exits the bag, it creates a pressure wave in the air with the same characteristic frequency (if the sound wave inside is hitting the wall of the bag 1000 times per second, the bag will vibrate a little 1000 times per second), and hence you hear a higher pitched note.

7.3 | ASSUMPTIONS OF THE KINETIC THEORY—A SECOND LOOK

7.3.1 Fluctuations from Equilibrium Values

The kinetic theory relied on converting the momentum transfer from individual collisions (which are very abrupt) into an average pressure. This will only be valid if the pressure we observe is the average of many events on an everyday timescale—in which case the fluctuations are small. This is a reasonable approximation, as we can illustrate by an example which might reflect an attempt to measure these fluctuations.

For example, suppose we measure the pressure with a simple U-tube manometer filled with mercury. Suppose the manometer is set up with a 1 cm diameter tube exposed to 1 atm nitrogen at room temperature (298K) on one end, and exposed to vacuum on the other end (which of course will be approximately 760 mm higher). The observed pressure can only change when the column of mercury has time to flow: the device (and any other measuring device) will have a nonzero response time. A reasonable estimate for the response time of a manometer might be 0.1 seconds, so the amount the pressure will appear to fluctuate will depend on the number of collisions with the top of the column in that time.

A pressure of 1 atm is approximately 10^5 Pa $= 10^5$ kg $\cdot$ m^{-1} $\cdot$ s^{-2} which is the force exerted by the gas per unit area by definition (Equation 7.3). The area of the top of the column is $\pi \times (0.005$ m$)^2 = 7.85 \times 10^{-5}$ m^2 so the average force on the column from the gas is

$$\begin{aligned} F &= P \cdot A = (10^5 \text{ kg} \cdot \text{m}^{-1} \cdot \text{s}^{-2})(7.85 \times 10^{-5} \text{ m}^2) \\ &= 7.85 \text{ kg} \cdot \text{m} \cdot \text{s}^{-2} \end{aligned}$$

which means that the amount of momentum which is transferred to the mercury column in $t = 0.1$ second is $F \cdot t = 0.785$ kg $\cdot$ m $\cdot$ s^{-1}.

The average collision generates a momentum transfer of $2\,m\,\langle|v_y|\rangle$ (Equation 7.7) which for N_2 is roughly

$$2\left(\frac{0.028\ \text{kg}\cdot\text{mol}^{-1}}{6.02\times 10^{23}\ \text{molecules}\cdot\text{mol}^{-1}}\right)(297\ \text{m}\cdot\text{s}^{-1}) = 2.76\times 10^{-23}\ \text{kg}\cdot\text{m}\cdot\text{s}^{-1}$$

So, in order to maintain this pressure, there must be $(0.785\ \text{kg}\cdot\text{m}\cdot\text{s}^{-1})/(2.76\times 10^{-23}\ \text{kg}\cdot\text{m}\cdot\text{s}^{-1}) = 2.8\times 10^{22}$ collisions in 0.1 second.

Finally, the rule of thumb given in Chapter 4 is that fluctuations in random processes scale roughly as the square root of the number of events. So the number of collisions will fluctuate by about $\sqrt{2.8\times 10^{22}} \approx 10^{11}$, and the pressure will fluctuate by about 1 part in 10^{11}—in other words, it will stay the same in the first ten or eleven digits of its value!

Realistically, fluctuations are larger than this because the atmosphere is not at equilibrium; both air currents and temperature variations will generate larger effects. But these examples show that the macroscopic average effect (the pressure) can be quite uniform, even though each molecule provides its contribution to the pressure only in the instant it collides with the walls, and thus the individual contributions are not at all uniform. Statistical averaging has dramatically simplified the apparent behavior.

7.3.2 Thermal Conductivity

Another assumption we made—that all collisions with the wall are elastic and transfer no energy because of the large mass difference—may strike you as quite strange if you think about it carefully. Heat transfer between gases and solids is readily observed. The air in your refrigerator will cool down food; the air in your oven will heat it.

Gas molecules do not actually bounce off the wall of a container (or your skin) as if it were a uniform massive structure, the way we sketched it in Figure 7.2: they collide with individual atoms at the wall surface, which are also moving because of vibrations. If the temperature of the wall and the gas are the same, on average the gas kinetic energy is as likely to increase or decrease as a result of any single collision. Thus a more accurate statement of the assumption required to derive the ideal gas law is that the walls and gas molecules are at the same temperature, so there is no average energy flow between the two.

If two plates with area a and separation d are maintained at temperatures T_1 and T_2, for a time t, the average heat flow q/t is given by

$$q/t = K(T_1 - T_2)a/d \tag{7.31}$$

K is called the **thermal conductivity**, which for air at STP is .023 W/(m $\cdot$ K). One way to reduce this energy flow is to decrease the pressure. Cryogenic liquids (such as liquid nitrogen, which boils at 77K) are commonly stored in **Dewar flasks**, which are double-walled containers with an evacuated region between the walls.

7.3.3 Collisions and Intermolecular Interactions

Very little had to be assumed to get to Equation 7.18—the most significant assumption was that the energy of the molecules could be written entirely as kinetic energy, with no potential energy. Thus we effectively pictured the molecules as infinitesimally small "hard spheres," which do not take up any space. This let us scale up Equation 7.10 by asserting that N molecules exert N times the pressure.

In reality, molecules each occupy some space, so the "empty" volume of the container decreases as the concentration N/V increases. In addition, there is generally some attraction even at distances substantially larger than the nominal diameter of the molecules, and the repulsive part is somewhat "soft" so that collisions are not instantaneous. The exact form of this interaction must be calculated by quantum mechanics, and it depends on a number of atomic and molecular properties as discussed in Chapter 3. For neutral, nonpolar molecules, a convenient approximate potential is the Lennard-Jones 6-12 potential, discussed in Chapter 3; Table 3.5 listed parameters for some common atoms and molecules.

$$U(r) = 4\varepsilon \left(\left(\frac{\sigma}{r} \right)^{12} - \left(\frac{\sigma}{r} \right)^{6} \right) \tag{7.32}$$

The most important difference between this potential and the hard-sphere potential is the addition of an attractive term at long distances. Thus molecules could stick together, just like you stick near the surface of the Earth. You remain on the Earth because the depth of the well generated by the Earth's gravitational field is larger than the kinetic energy you can achieve by jumping (see Problem 3-5). We refer to loosely bound pairs of molecules as **complexes**. For such a complex to survive, the total energy (kinetic plus potential) must be less than the potential energy as $r \rightarrow \infty$, just as it is for you on Earth.

We have defined the potential energy such that $U(r) \rightarrow 0$ as $r \rightarrow \infty$. Thus, with this definition of the zero of potential energy, forming a complex requires the total en-

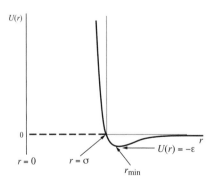

FIGURE 7.7 ▶ Comparison of a hard-sphere potential (dashed line) with a Lennard-Jones 6–12 potential (solid line).

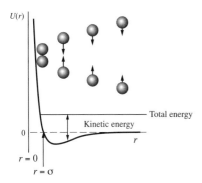

FIGURE 7.8 ▶ As the separation between two interacting molecules decreases, some of the total energy is converted into kinetic energy (in the region where $U(r) < 0$) and the molecules accelerate. Once they reach the point where $U(r)$ equals the total energy, the molecules must be motionless. Then they bounce back. If the total energy is positive, the two molecules cannot remain trapped in the potential well.

ergy to be negative. However, conservation of energy (as discussed in Section 3.1) can be used to prove that two molecules cannot stick after a collision. Figure 7.9 illustrates the case of two molecules with equal mass m viewed in a frame of reference moving with a velocity such that the total momentum $\vec{p} = m\vec{v}_1 + m\vec{v}_2 = 0$ (so $\vec{v}_1 = -\vec{v}_2$), but the result is quite general. Long before the collision, when the two molecules are separated by a large distance, the potential, kinetic, and total energies are:

$$U \approx 0 \ (\text{because } U(r) \to 0 \text{ as } r \to \infty)$$
$$K = \frac{m\,|\vec{v}_1|^2}{2} + \frac{m\,|\vec{v}_2|^2}{2} \qquad (7.33)$$
$$E = K + U \approx \frac{m\,|\vec{v}_1|^2}{2} + \frac{m\,|\vec{v}_2|^2}{2} > 0$$

As the molecules come closer $U(r)$ goes negative, and the kinetic energy must then increase to conserve the total energy. The molecules move towards one another more rapidly, but the total energy (kinetic plus potential) will be positive, hence sufficient at all times for the two molecules to eventually escape the attractive well.

Thus the only way to make a complex is to transfer some of the internal energy to another system. In practice, this means three or more molecules have to all be close enough to interact at the same time. The mean distance between molecules is approximately $(V/N)^{1/3}$ (the quantity V/N is the amount of space available for each molecule, and the cube root gives us an average dimension of this space). At STP 6.02×10^{23} gas molecules occupy ≈ 22.4 L $(.0224 \text{ m}^3)$ so $(V/N)^{1/3}$ is 3.7 nm—on the order of 10 molecular diameters. This is expected because the density of a gas at STP is typically a factor of 10^3 less than the density of a liquid or solid. So three-body collisions are rare. In addition, if the "well depth" $V(r_{\min})$ is not much greater than the average kinetic energy ($3k_B T/2$) molecules will not stick for long; even if two molecules combine, the next collision will probably blow the complex apart. At room temperature (300K), the average kinetic energy is 2×10^{-21} J, which is comparable to the well depth for all of the molecules listed in Table 3.5.

Nevertheless, the existence of an attractive part to the potential has important consequences, particularly at low temperatures or high densities. It prevents us from remaining in the gaseous state all the way down to absolute zero, because it permits condensation if the temperature is low enough. For an ideal gas, the energy $E = 3k_B T/2$ is independent of volume. For a real gas, since increasing the volume increases the average intermolecular distance (and thus changes the average potential energy), the internal energy actually depends on the volume as well as the temperature. This can permit cooling by expansion. To take an extreme case, a carbon dioxide fire extinguisher at room temperature sprays out CO_2 at its freezing point ($-78°C$)—which in some cases does more damage to equipment than the fire it is intended to extinguish. Air conditioners also work by expanding gas into a low pressure region (which becomes colder if the conditions and the gas are chosen correctly) and then recompressing the gas (which takes work, hence heats it) in a second region which is outside of the zone to be cooled.

▶ Mean Free Path and Mean Time Between Collisions

The ideal gas relation was derived under the assumption that each molecule travels undisturbed from wall to wall, which is certainly not true at common pressures and temperatures. To see this, we need to get some estimate of the *mean free path* λ (the mean distance a molecule travels before it undergoes a collision), and the mean time between collisions.

The mean separation between molecules is just $(V/N)^{1/3}$ as noted in the last section. However, the mean free path is not the same as the mean separation, because the molecule probably is not moving directly towards its nearest neighbor. λ depends on the *size* of the molecules as well. If we go back to this idea of molecules as hard spheres, big spheres are big targets and undergo collisions more often. A rigorous calculation requires some complicated algebra, but it is not hard to see how λ depends on physical parameters such as pressure or temperature. Calculating the expected functional form of some expression is often much easier than getting exact numerical values, and it provides a crucial "reality check."

If you double the concentration N/V ($= P/k_B T$), on average you expect to go only half as far before you encounter another molecule. We thus predict that λ is proportional to $V/N = k_B T/P$. If you double the "size of the target" (expressed as the collisional cross-section σ^2, where σ is the size in the hard-sphere potential), on average you will also only go half as far before you undergo a collision. Thus we also predict that λ is inversely proportional to σ^2.

These predictions are correct. The precise expression turns out to be:

$$\lambda = (3.1 \times 10^7 \text{ pm}^3 \cdot \text{atm} \cdot \text{K}^{-1}) \frac{T}{\sigma^2 P} \qquad (7.34)$$

for hard spheres. The expression is more complicated for a Lennard-Jones potential, but the functional dependences are the same. As an example, for N_2 we get $\lambda = 6.6 \times 10^4$ pm at 293K and 1 atm pressure, which is substantially larger than the mean separation.

We also sometimes evaluate the "mean time between collisions" τ, which is the mean free path λ divided by the mean speed $\left(\overline{s^2}\right)^{1/2}$. The reciprocal of τ gives the number of collisions per second, called Z:

$$\tau = \frac{\lambda}{\left(\overline{s^2}\right)^{1/2}}; \ Z = \tau^{-1} = \frac{\left(\overline{s^2}\right)^{1/2}}{\lambda} \tag{7.35}$$

For N_2 at 1 atm pressure and 293K, $\tau \approx 100$ ps.

▶ Diffusion

Intermolecular collisions do not cause large deviations from the ideal gas law at STP for molecules such as N_2 or He, which are well above their boiling points, but they do dramatically decrease the average *distance* molecules travel to a number which is far less than would be predicted from the average molecular speed. Collisions randomize the velocity vector many times in the nominal round trip time, leading to diffusional effects as discussed in Chapter 4. If all of the molecules start at time $t = 0$ at the position $x = 0$, the concentration distribution $C(x, t)$ at later times is a Gaussian:

$$C(x, t) \propto \exp(-x^2/\{4Dt\}) \tag{7.36}$$

where D is called the *diffusion constant* (technically the **self diffusion constant**, because we assumed all molecules are identical) and appears in many other problems as well. The functional dependence of D on pressure, temperature, mass, and molecular size can also be predicted by some simple substitutions. Let us go back to the analogy of the "coin toss" model for a random process, which we first made in Section 4.3. The number of "tosses" per second is analogous to the number of times the velocity gets randomized per second (the number of collisions per second, Z); the distance traveled between collisions, λ, is analogous to the length of the step we took after each toss in the earlier problem.

Doubling the step length λ would obviously double the distance you end up from the starting point (if the number of random events is constant) so $\langle x^2 \rangle^{1/2}$ is proportional to λ. Doubling the number of random events (if the step length is constant) would raise the expected deviation by $\sqrt{2}$, because of the nature of random processes as explained in Section 4.2, so $\langle x^2 \rangle^{1/2}$ is also proportional to $\sqrt{Z}$. So we would predict

$$\langle x^2 \rangle^{1/2} \propto \lambda \sqrt{Z}; \ D \propto \lambda^2 Z \tag{7.37}$$

We can combine the relations $\langle v_x^2 \rangle^{1/2} = \sqrt{kT/m}$, $\lambda \propto T/(\sigma^2 P)$ (Equation 7.34) and $\lambda Z = (s^2)^{1/2} \propto \sqrt{T/m}$ (Equation 7.35) to give

$$D \propto \lambda^2 Z = \lambda(\lambda Z) \propto \left\{ T/(\sigma^2 P) \right\} \left\{ \sqrt{T/m} \right\} = \left\{ \frac{T^{3/2}}{P\sigma^2 \sqrt{m}} \right\} \tag{7.38}$$

A detailed derivation would give the full expression, which is

$$D = (2/3)(k_B/\pi)^{3/2} \frac{(T)^{3/2}}{P\sigma^2 \sqrt{m}} \tag{7.39}$$

in accord with our predictions.

▶ Nonideal Gas Laws

The results of the last section showed that, for any macroscopic container at normal pressures, it is not reasonable to conclude that the molecules proceed from wall to wall without interruption. However, if the interaction potential energy between molecules at their mean separation is small compared to the kinetic energy, the speed distribution and the average concentration of gas molecules is about the same everywhere in the container. In this limit, the only real effect of collisions is the "excluded volume" occupied by the molecule, which effectively shrinks the size of the container. At 1 atm, only about 1/1000 of the space is occupied (remember the density ratio between gas and liquid), so each additional molecule sees only 99.9% of the container as free space. On the other hand, if the attractive part of the interaction potential cannot be totally neglected, the molecules which are very near the wall will be pulled slightly away from the wall by the other molecules. This tends to decrease the pressure.

Corrections to the ideal gas law can be introduced in many different ways. One well-known form is the ***van der Waals equation*** for a nonideal gas:

$$\left(P + \frac{a}{(V/n)^2} \right) \left(\frac{V}{n} - b \right) = RT \tag{7.40}$$

We can compare a quite nearly ideal gas (He, $a = 0.034$ L$^2 \cdot$ atm/mole^2, $b = .0237$ L/mole with a much less ideal gas (CO$_2$, $a = 3.59$ L$^2 \cdot$ atm/mole^2, $b = .0427$ L/mole). The b term reflects the excluded volume and does not change by much. The a term, reflecting intermolecular attractions, can change dramatically as the gas is changed.

The ideal gas law implies that at STP V/n is 22.4141 L/mole, so the b term leads to a correction of about 0.1% at STP. The a term at STP leads to a correction of about 1% for He. In fact, the experimental volume of one mole of He at STP is 22.434 L. The volume of CO$_2$ is 22.260 L, instead of the "ideal gas" value of 22.41410 L.

Another modified form of the ideal gas law is the ***virial expansion***:

$$\frac{PV}{nRT} = 1 + B(T) \left(\frac{n}{V} \right) + \cdots \tag{7.41}$$

This expansion in principle also includes terms proportional to $(n/V)^2$ and all higher powers of (n/V). However, when the density n/V is much smaller than the density of a solid or liquid, so that most of the container is empty space, this expansion converges rapidly and the higher terms can be ignored. $B(T)$ is called the *second virial coefficient*, and is a function of temperature.

The virial expansion has a far more solid theoretical justification than does the van der Waals equation. It can be shown quite generally that:

$$B(T) = -2\pi N_{\text{Avogadro}} \int_{r=0}^{r=\theta} (e^{-U(r)/k_B T} - 1)r^2 \, dr \tag{7.42}$$

If we completely ignore intermolecular interactions, $U(r) \approx 0$, $e^{-U(r)/k_B T} - 1 = 0$, and hence $B(T) = 0$. For a hard sphere potential, when $r > \sigma$, $U(r) = 0$, so again $e^{-U(r)/k_B T} - 1 \approx 0$; but when $r < \sigma$, $U(r)$ is infinite and $e^{-U(r)/k_B T} = 0$. Thus for hard spheres we can reduce Equation 7.42 to

$$B(T) = -2\pi N_{\text{Avogadro}} \int_{r=0}^{r=\sigma} (-1)r^2 \, dr = 2\pi N_{\text{Avogadro}} \frac{\sigma^3}{3} \tag{7.43}$$

If the distance of closest approach between two molecules is σ, the radius of each molecule must be $\sigma/2$, and the volume of each molecule would then be $4\pi(\sigma/2)^3/3 = \pi\sigma^3/6$. Thus B is four times the *excluded volume*, the part of the container occupied by the gas molecules themselves. The pressure is always higher than the ideal gas law would predict, and independent of temperature.

The integral in Equation 7.41 must be calculated by computer for a Lennard-Jones potential (Figure 7.9). This curve agrees remarkably well with experimentally mea-

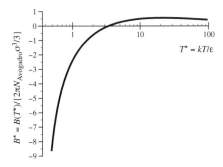

FIGURE 7.9 ▶ $B(T)$ for a Lennard-Jones 6-12 potential. B^* is the ratio between $B(T^*)$ and the hard-sphere value $2\pi N_{\text{Avogadro}}\sigma^3/3$. T^* is the ratio of the thermal energy $k_B T$ to the well depth $\mathcal{E}$. B^* and T^* make Figure 7.10 a "universal curve," valid for any gas. This curve agrees well with experiments on many molecules.

sured values of $B(T)$ for a large number of gases. As a result, second virial coefficients can be used to measure $\mathcal{E}$ and σ. Note that $B(T)$ is negative at low temperatures, positive at high temperatures, and levels off as the temperature gets extremely large. This is expected because the major effect of the attractive part of the potential—the acceleration of molecules as they grow closer—becomes less significant as the relative initial velocity increases, and this velocity is proportional to $\sqrt{T}$.

The temperature at which $B(T) = 0$ is the temperature at which a gas behaves most nearly like an ideal gas, and is called the **Boyle temperature** T_B. From Figure 7.9 we can see that

$$T_B \approx 3.4\mathcal{E}/k_B \qquad (7.44)$$

Finally, we should note that the a and b coefficients in the van der Waals equation are related to $B(T)$ in the virial expansion. In the limit that $b \ll V/n$, it can be shown with some algebraic manipulation (Problem 7-7) that

$$B(T) \approx b - \frac{a}{RT} \qquad (7.45)$$

7.4   SUMMARY

The kinetic theory of gases, coupled with the Boltzmann distribution, lets us predict a wide variety of the macroscopic properties of gases, and the agreement with experiment is excellent—even though we are making extremely crude approximations to the microscopic structure of the individual molecules. Molecules are much more complicated than tiny billiard balls, yet at standard temperature and pressure, the answers we get from the simplest theories are good to within a few percent.

▶ PROBLEMS ▶

7-1.* Solve Equations 7.5 and 7.6 algebraically to find the momentum transferred in the collision between a 1 kg wall and helium atom moving at 1000 m · s^{-1}.

7-2. Assume that a methane molecule hits a surface with impact velocity 250 m · s^{-1} (perpendicular to the surface). Assume further that the impact lasts for approximately 1 ps, and that the force is applied over an area of approximately 1 nm^2. Estimate the peak pressure at the point of impact.

7-3.* Suppose the density of a gas is kept constant, but the temperature is doubled. Predict what would happen to (a) the mean free path λ; (b) the mean time between collisions τ; (c) the diffusion constant D. If you can, use your intuition about the physical process, rather than substitution into equations derived in this chapter.

7-4. According to the ideal gas law, doubling the temperature while keeping the density N/V constant (for example, keeping the gas in a rigid container) doubles the pressure.

(a) For a nonideal gas (use the van der Waals equation), when the temperature is doubled at constant density, does the pressure exactly double? If not, is the increase in pressure larger or smaller than doubling?

(b) Illustrate the result in part (a) by calculating the pressure of one mole of helium in a 22.4 L container at 273K and 546K.

7-5.* Use the three-dimensional speed distribution to show that $s_{\text{most probable}} = \sqrt{2kT/m}$ and that $\langle s \rangle = \sqrt{8kT/\pi m}$.

7-6. Assume that you have a sample of uranium hexafluoride with the natural abundance (0.7%) of ^{235}U, and that you want to use gaseous effusion to separate the isotopes.

(a) What will be the concentration of ^{235}U in the gas after it effuses from the chamber?

(b) How many times must you repeat this process to produce 90% ^{235}U?

7-7.* We noted earlier that one mole of carbon dioxide, at STP (1 atm and 273K), occupies 22.260 L, instead of the "ideal gas" value of 22.41410 L. Use this result to calculate the second virial coefficient $B(273K)$ for carbon dioxide.

7-8. The diffusion constant for hydrogen at STP was given in Chapter 4 ($D = 1.5 \times 10^{-4}$ m$^2 \cdot$s^{-1}). Use this information to predict:

(a) the diffusion constant at 0.1 atm and 273K

(b) the hard-sphere cross section σ

(c) the mean free path between collisions at STP

(d) the mean time between collisions

7-9. Repeat the calculations in Problem 7-8 for carbon dioxide ($D = 1.0 \times 10^{-5}$ m$^2 \cdot$s^{-1}).

7-10. If we assume that $(V/n) \gg b$, then the term $((V/n) - b)$ in the van der Waals equation can be written as

$$\frac{V}{n} - b = \frac{V}{n}\left(1 - \frac{bn}{V}\right) \approx \frac{\frac{V}{n}}{1 + \frac{bn}{V}}$$

using the relation $1 - x \approx \frac{1}{1+x}$ for $x \ll 1$, Equation 1.41. Use this result to prove Equation 7.34.

7-11. Would your voice sound higher or lower in an 80% argon/20% oxygen mixture than in dry air (approximated as 80% nitrogen/20% oxygen)?

7-12. We went from the one-dimensional velocity distribution to the three-dimensional speed distribution by adding in an extra v^2 factor to account for the added degeneracy in "velocity space." The one-dimensional *diffusion* equation 7.36 has a form which is mathematically very similar to the one-dimensional velocity distribution (also a Gaussian). By analogy with the speed distribution, find an explicit equation for the rms displacement in any direction $\left(\overline{r^2}\right)^{1/2}$. (You can solve this without integrating!)

7-13. Suppose a gas in a piston expands slowly from pressure P_1 and volume V_1 to pressure P_2 and volume V_2, in the process doing work against the atmosphere. In many cases we can assume the expansion is *adiabatic*, meaning that no heat enters the gas during the expansion. If the heat capacity is independent of temperature, it can be shown that $P_1 V_1^{\gamma} = P_2 V_2^{\gamma}$, where the ratio $\gamma = c_p/c_v$ is the same quantity encountered in the speed-of-sound expression, Equation 7.29.

(a) If $V_2 > V_1$, does the gas temperature increase or decrease in an adiabatic expansion? Why?

(b) Suppose $P_1 = 2$ atm, $V_1 = 10$ L, the initial temperature is 273K, and the gas is adiabatically expanded until it reaches 1 atm. What will the final volume and temperature be if the gas is helium? What will they be is the gas is nitrogen?

7-14. Suppose we have a beam of light at $\lambda = 248$ nm (which turns out to be a convenient wavelength for a high-power laser). This laser pulse has 1 J of energy, lasts 10^{-12} s, and can be focuses to a spot with diameter 10 μm.

Assume this laser pulse is completely absorbed by a black wall. Use the relation $E = cp$ to calculate the momentum it transfers to the wall. By analogy with the calculations we did to derive the ideal gas law, calculate the "radiation pressure" the pulse exerts on the 10 μm spot while it is on.

Chapter 8

The Interaction of Radiation with Matter

If you want to understand function, study structure.

Francis H. C. Crick (1916–)
Co-discoverer of the double
helix structure of DNA.

Controlled radiation sources provide the most important modern tools for studying molecular structure and chemical dynamics. Virtually everything we know about the ways atoms interact has been deduced or confirmed by irradiation at a wide variety of wavelengths, from radiowaves to X-rays. In fact, protein and DNA structural determinations were the most important driving force in creating the modern chemical and molecular basis for the biological sciences.

Two important applications of radiation to determine molecular structure—X-ray crystallography and magnetic resonance—were discussed in Chapters 3 and 5. In this chapter we will discuss a variety of other techniques. Microwave absorption usually forces molecules to rotate more rapidly, and the frequencies of these absorptions provide a direct measure of bond distances. Individual bonds in a molecule can vibrate, as discussed classically in Chapter 3. Here we will do the quantum description, which explains why the *greenhouse effect*, which overheats the atmosphere of Venus and may be starting to affect the Earth's climate, is a direct result of infrared radiation inducing vibrations in molecules such as carbon dioxide.

Molecular absorption of visible or ultraviolet light usually excites electrons into higher energy states. Chemicals used as *dyes* absorb only a portion of the visible spectrum, and thus appear colored. Many of these dyes also generate *spontaneous emission* after their absorption, giving off a photon with somewhat lower energy. In fact, laundry detergents usually contain dyes that absorb ultraviolet light and glow in the visible; such dyes make white clothes look brighter in sunlight. Color comes from many other effects as well, as we discuss in Section 8.3. For example, the atmosphere transmits most of the visible radiation from the Sun, but scatters enough light to color the sky blue. The primary constituents of the atmosphere (oxygen and nitrogen) also transmit ultraviolet photons that are energetic enough to damage cells; ozone depletion in the upper atmosphere is dangerous because ozone absorbs some of this light.

Absorption is an intuitively reasonable process. If you have a large number of photons with the right energy, it makes sense that molecules can absorb some of those photons, and in the process move from a lower state L to an excited state U, in accord with Bohr's relation

$$\hbar\omega = (E_U - E_L) \tag{8.1}$$

The reverse of this process—molecules in an excited state dropping down to the ground state, and in the process amplifying a light field—is called *stimulated emission* and is the critical process involved in making a laser. Einstein predicted the existence of stimulated emission in 1916, but the first lasers were not built until 1960. The reason for the difficulty is Boltzmann's distribution (Chapter 4), which implies that under ordinary circumstances (e.g., at equilibrium) the lower state is always more populated than the higher state. So ordinarily, absorption is much more prominent than stimulated emission. Nonetheless, today there are dozens of different types of lasers, with wavelengths ranging from the microwave ("masers") all the way up to the X-ray region.

8.1 | INTRODUCTION TO ABSORPTION AND EMISSION

8.1.1 Absorption and Superposition States in Hydrogen Atoms

As discussed in Chapter 3, light and other radiation sources produce *electromagnetic fields* (electric and magnetic fields along perpendicular axes, oscillating between positive and negative values). In the next few sections, we will only consider the electric field, and we will begin by considering a very simple system: a hydrogen atom in its ground ($1s$) state. A constant electric field would exert a force on the electron, and an oppositely directed force on the proton (Figure 8.1). Because of the great mass difference, the electron is accelerated far more than is the proton. The net effect is to produce a charge distribution with the center of mass of the electron separated from the proton by an amount δ (the electric field induces a dipole moment).

Now imagine that the electric field were instantaneously turned off. The electronic cloud is distorted from the lowest energy state $\delta = 0$, so there will be a restoring force. It would be difficult to calculate the exact form of the potential energy, but we know that $\delta = 0$ gives the minimum energy, so the derivative vanishes there ($U'(0) = 0$). We can do a Taylor expansion about the $\delta = 0$ potential energy minimum as we did in

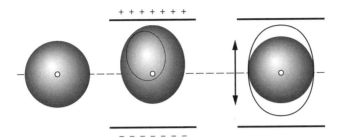

FIGURE 8.1 ▶ **Left**: a hydrogen atom in the ground $1s$ state. The center of the electronic cloud is on top of the nucleus, so there is no net dipole moment (average charge separation). **Center**: an electric field would push the electron and proton in opposite directions, creating a dipole moment proportional to the distance δ between the proton and the center of the electronic charge distribution. **Right**: if the electric field were then removed, the value of δ would oscillate with time.

Chapter 3. This predicts that for a small enough distortion, the potential energy function looks essentially like that of a harmonic oscillator:

$$U(\delta) \approx U(0) + U''(0)\frac{\delta^2}{2}$$

Thus the electron cloud will oscillate back and forth, just like a mass on a spring, or a child on a swing.

   Suppose we want to make a large displacement. One way to do this would be to apply an electric field which is strong enough to compete with Coulombic attraction over the 50 pm radius of a hydrogen atom. Unfortunately, this corresponds to an electric field strength of about 10^{11} V · m^{-1}, which is only achievable with extremely large laser systems. A simpler approach would be to take advantage of **resonance**. Driving the atom with an electromagnetic field which oscillates at the *same frequency* as the perturbed electron cloud will make the displacement grow larger with each cycle—just as periodically pushing a child on a swing eventually builds up a large motion with little effort.

   To find the frequency of oscillation for the perturbed electron cloud, recall that *any* hydrogen wavefunction can be written as some superposition of the stationary states. The stationary states by themselves have no dipole moment, since each of the hydrogen stationary states is centered at the nucleus. However, combinations of the stationary states can have a dipole moment. In particular, Figure 6.8 from the discussion of hybrid orbitals (Section 6.4.3) shows that combinations of s and p orbitals produce a charge separation (Figure 8.2). In fact, starting from an s state, the only way to produce a charge separation in the z-direction is by adding in a p_z state. So the perturbed orbital is a combination of $1s$ with a variety of p_z states.

   As discussed in Section 6.3, a superposition of two states gives a probability distribution which oscillates at frequency $\omega = (E_U - E_L)/\hbar$, and this is the same frequency

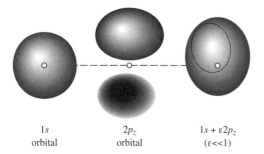

| 1s | $2p_z$ | $1s + \varepsilon 2p_z$ |
| orbital | orbital | ($\varepsilon \ll 1$) |

FIGURE 8.2 ▶ Both $1s$ and $2p$ orbitals give symmetric probability distributions. Combining the two into a superposition state breaks this symmetry, because the $2p$ orbital constructively interferes with $1s$ in one lobe and destructively interferes in the other.

as a photon that matches the energy difference between the two states (Equation 8.1). Every time one photon is *absorbed*, one atom moves to the excited state. We can express this process by the equation

$$N\hbar\omega + L \rightarrow (N - 1)\hbar\omega + U \tag{8.2}$$

which says that N photons (each with energy $\hbar\omega$) interact with one molecule, which is raised from L to U as the light weakens by one photon.

8.1.2 Selection Rules for Hydrogen Absorption

A light source with its electric field in the z-direction (z-polarized light) will excite hydrogen atoms from the ground $1s$ state ($n = 1, l = 0, m_l = 0$) to one of the states np_z (n arbitrary, $l = 1, m_l = 0$) *if* the energy of the photons matches the energy difference from $1s$ to the desired state. The frequencies in the **absorption spectrum** reflect the energy differences in these **allowed transitions**. Starting from $1s$, a single photon of z-polarized light will essentially never cause absorption to $2s$, or $3d$, or $110g$, or any state other than np_z.

This is the first example we will encounter of **selection rules** for allowed transitions. Physically, the selection rules arise because the electric field needs a dipole moment in order to interact with the atom, and only these specific changes in the quantum numbers create a dipole moment. More generally, the selection rules for absorption in a hydrogen atom are

$$\text{Absorption,} \atop z\text{-polarized radiation} \quad : \Delta n > 0, \Delta l = \pm 1, \Delta m_l = 0, \Delta m_s = 0 \tag{8.3}$$

There is nothing special about the choice of the z-axis for a hydrogen atom. If we had used x-polarized radiation, we would have excited np_x states; y-polarized radiation would have excited np_y states. Either of these are combinations of $m_l = +1$ and

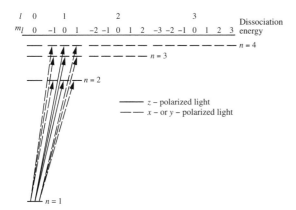

FIGURE 8.3 ▶ Allowed absorptions from the $1s$ state for hydrogen (levels with $n > 4$ omitted for clarity) which satisfy the selection rules presented in Equations 8.3 and 8.4.

$m_l = -1$. So with unpolarized radiation (field coming from all directions), the selection rules become

$$\text{Absorption, unpolarized radiation} \quad : \Delta n > 0,\ \Delta l = \pm 1,\ \Delta m_l = 0, \pm 1,\ \Delta m_s = 0 \qquad (8.4)$$

8.1.3 Spontaneous Emission

If we start with excited molecules or atoms, we can observe *spontaneous emission*—generation of light as the atom or molecule drops down to a lower energy level. Equation 8.1 also dictates the possible frequencies in the ***emission spectrum***. Spontaneous emission in general gives radiation in all directions, and the selection rules are almost the same as in Equation 8.4. The only difference is that we must have $\Delta n < 0$ for the final state to be lower in energy than the initial state.

$$\text{Spontaneous emission} \quad : \Delta n < 0,\ \Delta l = \pm 1,\ \Delta m_l = 0, \pm 1,\ \Delta m_s = 0 \qquad (8.5)$$

A hydrogen atom with its electron in a $2p$ orbital will decay back down to the $1s$ orbitral in approximately 1 nanosecond, giving off a photon with $\lambda = 121$ nm (determined by the energy difference between the two states). On the other hand, an electron in a $2s$ state is "stuck" (we call $2s$ a ***metastable level***), since emission to the only lower state ($1s$) is forbidden by the $\Delta l = \pm 1$ selection rule. On average, it takes about 100 ms for the electron to get back down to the ground state from $2s$.

8.1.4 Lasers and Stimulated Emission

In *spontaneous emission* the emitted light goes in a wide variety of directions. If two different atoms start from the same energy level and end up in the same level, they each produce photons at the same frequency; however, there is generally no correlation between the directions of these photons, or the phases of the two electric fields.

If a large number of photons are initially present at the right frequency, a fundamentally different process can be observed—*stimulated emission*. Here the existing light can "force down" additional molecules out of the excited state, and the newly created photons reinforce the existing light. Thus the material amplifies the light, which leads to the acronym "laser"—for "light amplification by stimulated emission of radiation." Modern laser applications range all the way from eye surgery to optical communications to laser fusion.

At high intensities, stimulated emission can be much stronger than spontaneous emission, and the important competition is between absorption and stimulated emission (Figure 8.4). If the population of the lower state is greater than the population of the upper state, more photons are absorbed than emitted, and the energy of the light beam decreases. If the population of the upper state is greater (which we call a ***population inversion***), the energy of the light beam increases.

Hydrogen and other one-electron atoms can be made into lasers because the state lifetimes vary so greatly. For example, an X-ray laser can be built by blasting carbon rods with an intense field, stripping off all the electrons. When the first electrons recombine with the nuclei, one-electron C^{5+} atoms are created in a wide variety of stationary states. Any population in the $2p$ states rapidly decays to the ground state; population in $3s$ or $3d$ decays more slowly. Thus the $3s \rightarrow 2p$ and $3d \rightarrow 2p$ transitions develop an inverted population distribution, and lase at the energy difference between the two states ($\lambda = 13.6$ nm).

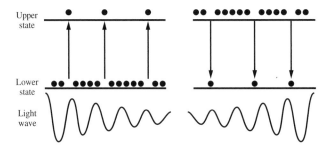

FIGURE 8.4 ▶ If more population is in the lower state than the upper state (left), absorption is stronger than stimulated emission, and radiation is absorbed from an external field. If the upper state is more populated (right), which never happens at equilibrium because of the Boltzmann distribution, an external field will be amplified. The situation on the right will create a laser.

8.2 MOLECULAR SPECTROSCOPY

This section extends the basic spectroscopic concepts we developed in Section 8.1 from hydrogen atoms to molecules. As we noted in Chapter 6, we cannot completely solve Schrödinger's equations for anything more complicated than a hydrogen atom, except by computer. However, many properties of the spectra of the simplest molecules (diatomic molecules) are quite simple and very useful.

The classical description in Chapter 2 separated molecular motions into translations, rotations, and vibrations. Each of these motions is treated differently in a quantum mechanical picture. In addition, electrons in molecules can be moved to higher energy levels, just as electrons in a hydrogen atom had multiple energy levels. We will treat each of these cases in turn.

8.2.1 Translational Energy

As we showed in Section 6.2, the stationary states for a particle in a box have specific values for kinetic (translational) energy. For a one-dimensional box, we showed that $E_n = n^2h^2/8mL^2$. For a three-dimensional box (side length L, volume $V = L^3$) each of the three degrees of freedom (motion in the x-, y-, and z-directions) has its own quantum number, usually called n_x, n_y, and n_z respectively. The energy is given by

$$E(n_x, n_y, n_z) = \frac{(n_x^2 + n_y^2 + n_z^2)h^2}{8mL^2} = \frac{(n_x^2 + n_y^2 + n_z^2)/h^2}{8mV^{2/3}} \tag{8.6}$$

For a macroscopic box (say $L = 0.1$ m) and realistic molecular masses, the separation between these levels is far less than $k_B T$. As a result, the distribution of energy levels appears virtually continuous, and quantum corrections to (for example) the ideal gas law are generally extremely small. However, it is possible to use Equation 8.6 to derive the ideal gas law from quantum mechanics instead of using the kinetic theory of gases (see Problem 8-10).

8.2.2 Rotational Energy

If a molecule rotates, it must have angular momentum. But one of the most important results of Chapter 6 was the observation that *angular momentum is quantized*. For a hydrogen atom, we showed that the allowed orbitals satisfied the conditions

$$\left|\vec{L}\right|^2 = \hbar^2 l(l + 1); l = 0, 1, 2, \ldots (n - 1)$$
$$L_z = \hbar m_l, m_l = -l, -l + 1, \ldots l - 1, l$$

This reflects the orbital motion of the electron around the nucleus.

For a diatomic molecule, the orbital motion of the two nuclei about each other (the rotational motion) satisfies *exactly the same equations*, except that the upper limit imposed on l by the principal quantum number n does not apply. By convention, we write the angular momentum of a molecule as J, to distinguish it from the atomic angular momentum L:

$$
\begin{aligned}
\left|\vec{J}\right|^2 &= \hbar^2 j(j+1); \, j = 0, 1, 2, \ldots \\
J_z &= \hbar m_j; \, m_j = -j, -j+1, \ldots j - 1, j
\end{aligned} \tag{8.7}
$$

Furthermore, the wavefunction describing the bond direction for quantum numbers j and m_j is *identical* to the angular portion of the hydrogen orbitals for quantum numbers l and m_l. So the ground state $j = m_j = 0$ is spherically symmetric, just like an s orbital; the $j = 1$ states look like p orbitals, and so on.

We showed in Chapter 6 that for a classical orbit $E = \left|\vec{J}\right|^2 / 2I$, where $I = \mu r^2$ and $\mu = m_1 m_2 / (m_1 + m_2)$ are the moment of inertia and the reduced mass respectively. Combining these relations with Equation 8.7 gives

$$
E = \frac{h^2 j(j+1)}{2\mu r^2}; \, \mu = \frac{m_1 m_2}{m_1 + m_2} \tag{8.8}
$$

Notice that the lowest energy state $j = 0$ has $E = 0$, but it does *not* correspond to a bond which is merely pointing in one direction in space; just like an s orbital, it is simultaneously pointing in all directions. This is yet another manifestation of the Uncertainty Principle. If a bond is known to point in one direction, the positional uncertainty perpendicular to the bond direction is zero, and the momentum uncertainty is infinite!

Any molecule with a permanent electric dipole moment can interact with an electromagnetic field and increase its rotational energy by absorbing photons. Measuring the separation between rotational levels (for example, by applying a microwave field which can cause transitions between states with different values of j) let us *measure the bond length*. The selection rule is $\Delta j = +1$—the rotational quantum number can only increase by one. So the allowed transition energies are

$$
\begin{aligned}
E(j = 1) - E(j = 0) &= \frac{\hbar^2}{\mu r^2} \\
E(j = 2) - E(j = 1) &= \frac{2\hbar^2}{\mu r^2} \\
E(j = 3) - E(j = 2) &= \frac{3\hbar^2}{\mu r^2}
\end{aligned} \tag{8.9}
$$

and so forth.

Rotational transitions are found in the microwave region of the electromagnetic spectrum. For example, $^{12}C^{16}O$ absorbs microwave radiation at $\nu = 115.27$ GHz and at multiples of this frequency, which corresponds to an absorbed photon energy of $E = h\nu = 7.6380 \times 10^{-23}$ J, and a wavelength of $\lambda = c/\nu = .0026$ m. Since the masses of the different isotopes are known to high accuracy, we can use this number to determine the bond length. The reduced mass μ for $^{12}C^{16}O$ is 1.1385×10^{-26} kg (see Section 3.5). The energy difference between the two lowest states ($\hbar^2/\mu r^2$ from Equation 8.9 above) then becomes

$$7.6380 \times 10^{-23} \text{ J} = \frac{(1.05457 \times 10^{-34} \text{ J} \cdot \text{s})^2}{(1.1385 \times 10^{-26} \text{ kg})r^2}$$

which gives $r = 112.8$ pm as the bond length in $^{12}C^{16}O$. Different isotopes of carbon or oxygen will give the same bond length, but the frequency of the absorbed radiation changes because μ changes (Problem 8-11).

For historical reasons, the energy differences between rotational levels are usually characterized by the quantity $\tilde{\nu} = E/hc$, with c expressed in centimeters per second ($c = 2.99793238 \times 10^{10}$ cm $\cdot$ s^{-1}). So most textbooks will write:

$$\tilde{\nu} = \frac{E}{hc} = \frac{\hbar j(j+1)}{8\pi c\mu r^2} = Bj(j+1); \quad B = \hbar/8\pi c\mu r^2 \quad (8.10)$$

Both $\tilde{\nu}$ and B have units of (1/cm). You should verify that B for $^{12}C^{16}O$ is 1.9225 cm^{-1}, and that $\tilde{\nu}$ for the lowest frequency transition in $^{12}C^{16}O$ is 3.845 cm^{-1}. For light, $\tilde{\nu} = 1/\lambda$, the reciprocal of the wavelength ($\lambda = 0.26$ cm $= .0026$ m as described above). $\tilde{\nu} = 1$ cm^{-1} corresponds to a wavelength of 1 cm, or a frequency of 29.979 GHz.

For most molecules, the energies of the lowest rotational states are substantially less than $k_B T$ at room temperature, so the Boltzmann distribution implies that many rotational levels are populated (Problem 8-6). In fact, since states with different values of m_j are degenerate, the ground state $j = 0$ is only the most populated state at very low temperatures. The relative population of different levels is given by

$$\frac{N(j)}{N(j=0)} = (2j+1)e^{-hcBj(j+1)/k_B T} \quad (8.11)$$

where the $(2j+1)$ factor arises because each value of j gives $(2j+1)$ different m_j states with the same energy. A convenient quantity to remember is the value of $k_B T/hc$ at room temperature (300K), which is 208 cm^{-1}. Thus for CO at room temperature

$$\frac{N(j=1)}{N(j=0)} = 3e^{-2B/(k_B T/hc)} = 3e^{-2\times 1.9225/208} = 2.94$$

This kind of *microwave spectroscopy* is the best technique available for determining the structure of small molecules in the gas phase. Microwave frequencies can be measured with extremely high accuracy, permitting bond length measurements literally to four or five significant digits. All of the bond lengths in Table 3.2 were found from the experimentally measured rotational frequency.

Polyatomic molecules have more complex microwave spectra, but the basic principle is the same; any molecule with a dipole moment can absorb microwave radiation. This means, for example, that the only important absorber of microwaves in the air is water (as scientists discovered while developing radar systems during World War II). In fact, microwave spectroscopy became a major field of research after that war, because military requirements had dramatically improved the available technology for microwave generation and detection. A more prosaic use of microwave absorption of water is the *microwave oven*; it works by exciting water rotations, and the tumbling then heats all other components of food.

8.2.3 Vibrational Motion

For our classical model of a molecule (two balls connected by a spring), the potential energy is given by $U(r) = k(r - r_e)^2/2$, where k is the force constant and r_e is the position of the potential minimum (the rest length of the spring), and oscillation occurred at frequency $\omega = \sqrt{k/\mu}$.

The potential energy function for a chemical bond is far more complex than a harmonic potential at high energies, as discussed in Chapter 3. However, near the bottom of the well, the potential does not look much different from the potential for a harmonic oscillator; we can then define an effective "force constant" for the chemical bond. This turns out to be another problem that can be solved exactly by Schrödinger's equation. Vibrational energy is also quantized; the correct formula for the allowed energies of a harmonic oscillator turns out to be:

$$\frac{E(v)}{hc} = \tilde{\nu} = \omega_e \left(v + \frac{1}{2} \right), \quad \omega_e = \frac{1}{2\pi c}\sqrt{k/\mu}, \quad v = 0, 1, 2, \dots \tag{8.12}$$

as graphed in Figure 8.5. Note that ω_e also has units of (1/cm). We introduced ω_e in Equation 8.12 for the same reason we introduced B in Equation 8.10—consistency with conventional usage—although it presents potential for confusion with the classical vibrational frequency $\omega = \sqrt{k/\mu}$ which is written in radians per second. We thus have the (exceedingly ugly) relationship

$$\omega_e = \frac{\omega}{2\pi \cdot \left(2.9979 \times 10^{10} \text{ cm} \cdot \text{s}^{-1} \right)} \tag{8.13}$$

For example, molecules of $^{12}C^{16}O$ can be excited from $v = 0$ to $v = 1$ by a photon with energy 4.257×10^{-20} J ($\omega = E/\hbar = 4.037 \times 10^{14}$ radians per second; $\omega_e = 2140$ cm^{-1}). This energy difference corresponds to the infrared region of the spectrum. This means the force constant for the C = O bond is $k = 1855$ N $\cdot$ m^{-1}. All of the force constants in Table 3.2 were found from the experimentally measured vibrational frequency.

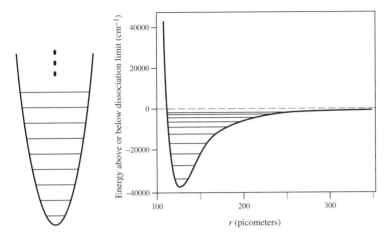

FIGURE 8.5 ▶ **Left**: A harmonic oscillator (potential energy $U(x) = kx^2/2$) has equally spaced energy levels, separated by the same energy as a photon with the classical vibrational frequency $\sqrt{k/\mu}$. The energy separation between adjacent levels is independent of v. The lowest state has $E > 0$; the system cannot be motionless. **Right**: For a realistic potential energy function, the energy levels grow closer as v increases. The first ten energy levels for $H^{35}Cl$ are shown here.

The lowest lying level (v = 0) has $E = hc\tilde{v}/2 = \hbar\omega/2$, not zero. In Problem 6.11 we showed that the wavefunction for the lowest state is

$$\Psi_{v=0} = Ce^{-x^2\sqrt{mk}/2\hbar} \tag{8.14}$$

which is a Gaussian. This implies that the atoms do not sit at a fixed separation (Problem 8-9). It is impossible for the two masses to simply sit at their equilibrium distance, even at absolute zero. This is another consequence of the Uncertainly Principle—motionless masses would imply exact knowledge of the distance between them, hence an infinite uncertainty in the momentum. Notice also that the energy separation between adjacent levels $E(v + 1) - E(v) = \hbar\omega$ is independent of v, and is the same as the energy of a photon with the classical vibrational frequency ω.

For any real molecule the potential looks significantly different from a harmonic oscillator at high energies, and the spacing decreases. Measurement of the energies of a large number of vibrational levels permits calculation of the actual potential energy function $U(r)$ and this has been done for many diatomic molecules (Figure 8.5) .

A harmonic oscillator can only change its vibrational quantum number by one when it absorbs a photon ($\Delta v = 1$); therefore, the only frequencies which can be absorbed are near the classical vibrational frequency $\omega = \sqrt{k/\mu}$. The absorption will also change the rotational quantum number ($\Delta j = \pm 1$). In practice, this means that the infrared spectrum of a small molecule has rotational structure, which permits bond length measurement as well as force constant measurement (Figure 8.6).

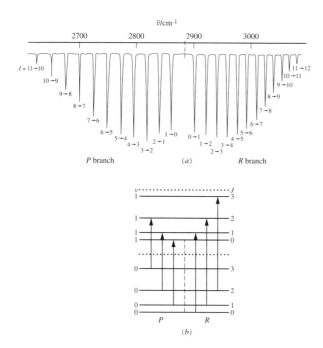

P branch   (a)   R branch

(b)

FIGURE 8.6 ▶ High-resolution infrared spectrum of HCl gas. All the transitions here have
$\Delta v = +1$, but the lines divide into two "branches," corresponding to $\Delta j = +1$ or $\Delta j = -1$. A few
lines are labeled. Note that each line is actually slightly split, because the separation between
rotational states is difference for $H^{35}Cl$ (75% natural abundance) and $H^{37}Cl$ (24% natural
abundance). The force constant and the bond length can both be extracted from the line positions.

A molecule can only absorb infrared radiation if the vibration changes the dipole
moment. Homonuclear diatomic molecules (such as N_2) have no dipole moment no
matter how much the atoms are separated, so they have no infrared spectra, just as they
had no microwave spectra. They still have rotational and vibrational energy levels; it
is just that absorption of one infrared or microwave photon will not excite transitions
between those levels. Heteronuclear diatomics (such as CO or HCl) absorb infrared radiation. All polyatomic molecules (three or more atoms) also absorb infrared radiation,
because there are always some vibrations which create a dipole moment. For example,
the "bending" modes of carbon dioxide make the molecule nonlinear and create a dipole
moment, hence CO_2 can absorb infrared radiation.

The vibrational frequency depends on the reduced mass and the force constant.
Often the individual bonds in polyatomic molecules generate vibrational frequencies
which depend only slightly on the rest of the molecule. For example, carbon-oxygen
double bonds are found in a wide variety of organic molecules (such as acetone,
$(CH_3)_2C=O$). The C=O "stretch" is excited at $\tilde{\nu} = 1750$ cm^{-1} in virtually any such
molecule, and is often used to confirm the existence of a C=O group in an unknown

sample. Other bonds have quite different frequencies; for example, the C–H stretch in methanol (H_3C–O–H) is observed around 3000 cm^{-1}. Thus infrared spectra are commonly used to extract structural information about complex molecules.

Vibrational excitations play an enormously important role in the world around us. For example:

1. Water is blue because of vibrations. The vibrational frequencies of the three normal modes of water are 3657, 3756 and 1595 cm^{-1} respectively (see Figure 3.13; recall that 1 cm^{-1} is 29.979 GHz). Since water is not a perfect harmonic oscillator, transitions with $\Delta v > 1$ are possible, although very weak. Transitions with $\Delta v \approx 6$ of the two stretching modes are in the visible region of the spectrum. These transitions are stronger in the red than in the blue. Hence the light which travels through many meters of water (and then gets scattered) is dominantly blue.

2. As discussed in Chapter 6, most of the Sun's energy reaches the Earth in the form of visible radiation, which is not absorbed by any of the major constituents of the atmosphere. However, the Earth has a mean surface temperature of approximately 290K, so the peak of *its* radiated energy is at $\lambda_{max} \approx 10$ μm (see Chapter 5), or $\tilde{v} \approx 1000$ cm^{-1}. Thus the most important mechanism for the Earth's cooling is emission of infrared radiation into space.

Absorption or reflection of infrared radiation serves to keep in heat. For example, the interior of a car or a greenhouse can get quite warm on a sunny day. This is called the *greenhouse effect*, and arises because glass is transparent to sunlight but absorbs and reflects much of the infrared radiation. In fact, semiconductor coatings for windows have recently become commercially available; these coatings are transparent in the visible but reflect still more infrared radiation, and dramatically decrease heat loss (hence energy consumption) in homes.

Infrared absorption in the atmosphere can have the same effect. Over the last century the concentration of carbon dioxide in the atmosphere has risen dramatically because of combustion. As a result, the atmosphere now absorbs more infrared radiation than it did in the past, and cooling into space is less efficient. A likely consequence is *global warming* , although a detailed calculation of the magnitude of the expected effect is far from simple. For example, while is it not difficult to estimate total CO_2 emissions from combustion, most of these molecules end up in the ocean as carbonates or bicarbonates, and do not directly contribute to global warming. Nonetheless, there is broad consensus in the scientific community that carbon dioxide emissions will tend to increase the Earth's temperature over the next few decades, with environmental consequences which may be severe.

8.2.4 Chemistry and the Origins of Color

Color—the selective absorption, emission, or reflection of components of the visible spectrum—can be produced by a variety of mechanisms. We already noted that water is blue because of vibrational excitation. The sky is blue because light gets scattered more strongly as the wavelength decreases (the scattering rate is proportional to λ^{-4}), hence the atmosphere is indirectly illuminated more by the blue components of sunlight than by the red components. Sunsets are red for the same reason; as the sun gets lower in the sky, the path that light travels through the atmosphere increases, and eventually the loss of blue light becomes noticeable. Rainbows are created because water droplets bend blue light more than red; peacock feathers are brilliantly colored because nature gave them diffraction grating on their feathers.

By far the predominant origin of color, however, is electronic excitation. We have already seen that the different orbitals of a hydrogen atom have different energies, and transitions between these orbitals (observed, for example, in a hydrogen discharge tube) can create colored light. The electrons in molecules lie in orbitals as well. Transitions from the molecular orbitals which are occupied at equilibrium to unoccupied orbitals can cause selective absorption of specific wavelengths; emission of light after the electrons have been excited can cause a colored "glow."

All molecules absorb light at short enough wavelengths. However, the energy difference between the *highest occupied molecular orbital* (HOMO) and the *lowest unoccupied molecular orbital* (LUMO) is difficult to calculate accurately for even small molecules. We can estimate this gap (and hence the absorption wavelength for electronic excitation) by a modification of the "particle-in-a-box" model considered in Chapter 6. We can view the Coulomb's law attraction between electrons and nuclei as confining the electrons (mass $m_e = 9.109 \times 10^{-31}$ kg) to a region comparable in size to a chemical bond ($L \approx 200$ pm). In that case, the formula for the energy levels of a particle-in-a-box (Equation 6.21) would give:

$$E_n = n^2 h^2 / 8mL^2 \approx n^2 (1.5 \times 10^{-18} \text{ J}) \tag{8.15}$$

The separation between the $n = 1$ and $n = 2$ levels is then 4.5×10^{-18} J, which is the energy of a photon with frequency $\nu = E/h = 6.79 \times 10^{15}$ Hz—well into the vacuum ultraviolet region of the spectrum. This suggests (and experiments confirm) that the vast majority of molecules do not undergo electronic excitation with visible light; the energy per photon is far too small. For example, oxygen and nitrogen are nearly transparent for $\lambda \geq 190$ nm.

Since the separation between energy levels is proportional to L, the minimum energy per photon needed for absorption decreases if the electron can be "delocalized" over several chemical bonds. A series of alternating single and double bonds (called *conjugated double bonds* creates exactly such a delocalization. Consider, for example, the molecule 11-cis-retinal (Figure 8.7). The term "cis" refers to a double bond with substituent groups on the same side of the bond, as shown by the arrow; a double bond with substituent groups on opposite sides of the bond is called "trans." The "11" just identifies the location of the bond. Notice that this molecule has a total of six double bonds (five trans, one cis), each one of which is created by p orbitals from two separate atoms. Thus twelve consecutive atoms have p orbitals which contribute to this chain

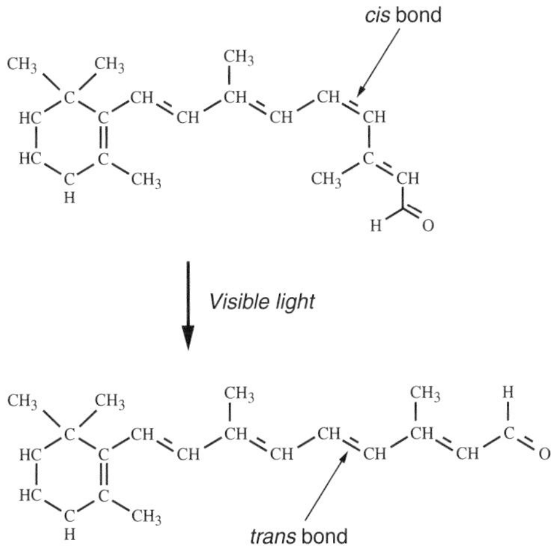

FIGURE 8.7 ▶ The central event in human vision is absorption of a visible photon by 11-cis-retinal, which then rearranges about one of its double bonds. The chain of alternating (conjugated) single and double bonds permits absorption in the visible.

(all of which are pointed out of the plane of the paper, as the molecule is drawn on the page). These orbitals together create a highly extended "box" which reduces the energy needed to excite an electron down into the range of visible light.

When this molecule absorbs light, it rapidly *isomerizes* (changes its structure) to the all-trans form. This specific chemical reaction is the central event in human vision. The molecule is bound within a protein (the combination is called *rhodopsin*); the isomerization triggers a series of later processes, ultimately leading to an electrical signal which is sent to the brain.

Virtually all dyes are organic molecules which get their colors either from a conjugated linear chain (like the one in retinal), from rings of conjugated bonds (Figure 8.8), or a combination of these structural features. This figure shows three simple molecules (benzene, anthracene and pentacene) which have progressively more rings fused together. We explicitly show here only one of several structures which can be drawn with alternating single and double bonds for each molecule. The actual structure is better represented as a superposition of these different structures, which are called ***resonance hybrids***. As the number of alternating single and double bonds increases, the separation between the energy levels decreases. This means that λ_{max}, the wavelength of the most intense absorption, increases as well.

Some, but not all, molecules emit light after they are excited. This is called *fluorescence* (if it happens quickly, nanoseconds to microseconds) or *phosphorescence* (if it takes a long time, milliseconds to minutes). Molecules generally fluoresce or phosphoresce at longer wavelengths than they absorb, because of energy conservation—a molecule can absorb a photon with energy $\hbar\omega_1$ and emit a photon with energy

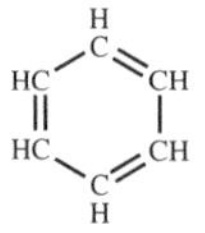

benzene (1_{max} = 200 nm)
colorless; no visible flourescence

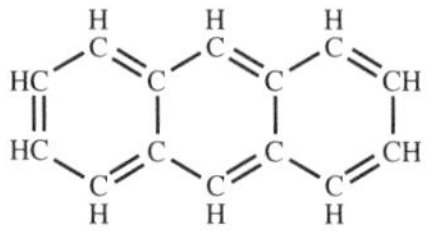

anthracene (λ_{max} = 376 nm)
colorless; violet flourescence

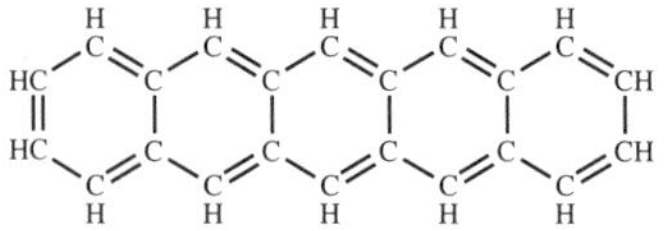

pentacene (λ_{max} = 575 nm)
blue; red flourescence

FIGURE 8.8 ▶ Comparison of the wavelength of maximum absorbance λ_{max}, the color, and the emission from three molecules in a series with 1, 3, and 5 fused aromatic rings respectively. The emission and absorption shift to longer wavelengths as the molecule grows.

$\hbar\omega_2 < \hbar\omega_1$, leaving $\hbar\omega_1 - \hbar\omega_2$ behind as rotational, vibrational, or electronic energy. Thus pentacene absorbs red light (making it appear blue) and gives off still redder fluorescence; anthracene absorbs ultraviolet light ($\lambda < 400$ nm) and gives off violet fluorescence. Phosphorescent compounds are used for "glow-in-the-dark" paints, which in effect store energy during the day and release it slowly at night. Such compounds often can also be made to glow by bombarding them with electrons, and are used for television screens.

At room temperature, all but the smallest molecules have very broad absorption and fluorescence spectra (typically absorption or fluorescence spectra have linewidths $\Delta\lambda \approx 100$ nm). Molecules start in a large number of different vibrational states (recall that the number of distinct vibrations for a N-atom nonlinear molecule is $3N - 6$), electronically excited molecules relax quickly (particularly in solution or solids), and many different final vibrational states are possible.

8.3 MODERN LASER SPECTROSCOPY

The invention of the laser in the late 1950s, and demonstration of the first practical laser in 1960, were immediately recognized as scientific accomplishments of the first magnitude, but for years the laser was derided as "a solution in search of a problem." Early sources were difficult to use and maintain, and the range of available wavelengths was quite limited. Massive efforts in laser source development eventually paved the way for important commercial applications. By 1994, the U.S. National Academy of Sciences estimated the overall annual economic impact of lasers and related products at $100 billion.

In the early 1990s a technological revolution occurred with the development of tunable solid-state lasers. As of this writing, the most important tunable laser system is the *titanium-doped sapphire laser*, which can be made to operate throughout the deep red and near infrared ($\lambda = 700$–1000 nm). At one extreme limit, continuous lasers can be made *monochromatic* (consisting of only a single wavelength or frequency) to extremely high precision; commercially available lasers have fractional bandwidths of one part in 10^9, and specialized lasers in laboratories can do at least a factor of 1000 better than this. Such lasers have an electric and magnetic field which looks very much like a single sine wave, oscillating for trillions of cycles with a single, well-defined period. Average powers of 1–10 W are readily produced.

Continuous lasers are important tools for measuring molecular spectra in stable molecules, which can then give information about structure. Consider, for example, the molecule I_2, which is a stable gas at room temperature. Suppose we wish to determine the structure of this molecule. The molecule vibrates and rotates, and as discussed earlier, measurement of the differences in energy between the different vibrational and rotational levels can give the bond length and force constant. Since I_2 has no dipole moment, it does not absorb infrared or microwave radiation. However, it does absorb visible light, which excites the molecule into a higher, normally empty electronic state. In the process, vibrational and rotational quantum numbers change, so the exact positions of the spectral lines reveal the vibrational and rotational energy level differences in both the excited and ground states. Measurement of the positions of these lines gives the complete vibrational potential energy function $U(r)$ in either state and the rotational constant ($B = .037$ cm^{-1} in the ground state, $B = .027$ cm^{-1} in the excited state).

Other types of lasers can generate extremely short pulses. As of this writing, the shortest laser pulses ever generated had a duration of 3.5 fs, although pulse lengths of 50–100 fs are more common. Such laser pulses are *not* monochromatic; they have a very large bandwidth (hence a broad distribution of wavelengths in a single pulse) from the uncertainty relation $\Delta \nu \Delta t \approx 1/4$ (Problem 5-12). However, they can concentrate an enormous amount of energy into a short burst. Commercially available laser systems give very short pulses as rapidly as once every 10 ns, or as rarely as once per second. Again the average power is generally in the 1–10 W range, so the peak power is vastly higher with low repetition rate. A common compromise between high pulse energy and high repetition rate would be to give one pulse per millisecond, with pulse energies of a few mJ.

The titanium-sapphire wavelength itself is too short for vibrational excitation in most molecules, and too long for electronic excitation. However, these high peak powers permit efficient *frequency conversion*. For example, certain crystals can convert two photons with frequency ω into a single photon with frequency 2ω. In many ways this can be viewed as similar to a second-order chemical reaction, such as the dimerization of NO_2 to form N_2O_4. The rate of that reaction is proportional to the square of the NO_2 concentration; the rate of this *frequency doubling* is proportional to the square of the "photon concentration" (the intensity), so high powers are very useful. It is also possible to combine two photons with different frequencies ω_1 and ω_2 in either *sum-*

frequency generation (creating a single photon with frequency $\omega_1 + \omega_2$) or ***parametric oscillation*** (destroying one photon with frequency ω_1 while creating two photons at lower frequency, one with frequency $\omega_1 - \omega_2$, the other with frequency ω_2). The upshot of all of this is that pulsed titanium-sapphire lasers can serve as the starting point to efficiently make laser pulses all the way from the deep ultraviolet to the far infrared.

For example, pairs of extremely short (***ultrafast***) laser pulses have characterized the dynamics of the cis-trans isomerization of rhodopsin (Figure 8.7). The first pulse of visible light (called a "pump" pulse), which lasts for 35 femtoseconds, puts the cis molecule into an electronically excited state. The second pulse (called a "probe" pulse) is only 10 femtoseconds long, so it contains a broad distribution of wavelengths. The absorption of this pulse at different wavelengths is measured, and the delay between the two pulses is varied. When the delay between the pulses is very short ($\ll 200$ fs) only the absorption spectrum of cis-11-retinal is seen. The absorption spectrum of trans-11-retinal, which is not initially present in the sample, is seen for delays longer than 200 fs. Thus, the researchers were able to conclude that the first chemical step of vision is one of the fastest photochemical reactions known, and is essentially complete in 200 fs. Such "pump-probe experiments" are widely used in chemical physics laboratories. The value of such "femtochemistry" experiments was recognized by the awarding of the 1999 Nobel Prize in Chemistry to Ahmed Zewail.

Lasers also have many research applications outside of chemistry. They can be modulated (turned on or off, or changed in frequency) in tens of femtoseconds, and this means that they can transmit many "bits" of information in a very short time. Intense laser beams can cut metal or human tissue with high precision. They can even generate high pressures (photons have momentum, so bouncing light off a surface exerts a pressure, just as bouncing gas molecules off a surface exerted pressure), and this is used to induce nuclear fusion.

► PROBLEMS ►

Note: Some of these problems require high resolution atomic masses, listed in Appendix A.

8-1. For the molecule ^{1}H^{35}Cl, the microwave spectrum consists of a series of equally spaced lines at $\tilde{v} = 21.18$ cm^{-1} and at $2\tilde{v}$, $3\tilde{v}$, ... Find the length of the H-Cl bond.

8-2. Equation 8.6 gives the energy levels for a particle in a three-dimensional box. Suppose $L = 1$ meter, and m corresponds to a small atom such as a helium atom. We treated this problem classically by the kinetic theory of gases in Chapter 5, which is equivalent to assuming that a continuous distribution of energies is allowed. To see if this approximation is valid, use the Boltzmann distribution to find the energy level which is most highly populated at $T = 300$K, and how large n has to be before you get to levels which are only half as populated at $T = 300$K as this most populated level.

8-3.⋆ The bond length in $^{127}I^{35}Cl$ is 232.1 pm. Find the frequency of light absorbed when the molecule makes a transition from $j = 0$ to $j = 1$.

8-4. Find the minimum vibrational energy for $^{23}Na^{35}Cl$, using the data in Table 3.2. Also find B and ω_e for this molecule.

8-5.⋆ Find B and ω_e for the nitrogen molecule, using the data in Table 3.2.

8-6. There are $(2j+1)m_j$ levels for each value of j, so the Boltzmann distribution tells us that the total number of molecules with rotational number j is given by:

$$\frac{n(j)}{n(0)} = (2j + 1) \exp(-E_j/kT)$$

For HCl, which rotational level is most populated at room temperature ($T = 300K$)?

8-7.⋆ Predict the color of the molecule tetracene, which has four fused rings and thus is intermediate in size between anthracene and pentacene. Also predict the color of the molecule's fluorescence.

8-8. As noted in Section 8.3, the rotational constant for I_2 is $B = .037$ cm^{-1} in the ground state and $B = .027$ cm^{-1} after excitation to the electronic state which makes the vapor purple. Calculate the change in bond length upon electronic excitation.

8-9.⋆ As noted in Chapter 3, we can use the work-energy theorem to describe the change in energy upon expansion of a gas. For a gas which is nearly at the same pressure as its surroundings, the work done by expansion is $\Delta w = P\Delta V$; by conservation of energy, this work must come from the internal energy of the gas, so $\Delta E = -P\Delta V$. In the limit of very small changes, this means $(dE/dV) = -P$. Combine this result with the expression for the particle in a box (Equation 8.6) to show that $PV = 2E/3$. (We have already shown from the Boltzmann distribution that the translational energy of one mole of gas is $E = 3RT/2$, so combining these results gives the ideal gas law without resorting to the kinetic theory.)

8-10. As noted in the text, the average bond length in a diatomic molecule can be determined to high precision by microwave measurements. However, even in the ground state the molecule still has total energy $E = \hbar\omega/2$, so it is still vibrating. Use the force constant for CO from Table 3.2 to predict the spread Δx in the probability distribution.

8-11.⋆ Predict the line positions (in cm^{-1}) in the rotational spectrum of $H^{127}I$.

8-12. Find the change in B and ω_e in carbon monoxide if the carbon isotope is carbon-13 instead of carbon-12.

8-13.⋆ What would be the color of a very deep pool of D_2O?

Appendix A

Fundamental Physical Constants

Name of Constant	Symbol	Value	Uncertainty (ppm)
Speed of light in vacuum	c	$2.997\,924\,58 \times 10^8$ m·s^{-1}	(exact)
Permittivity of free space	μ_0	$4\pi \times 10^{-7}$ m·kg·C^{-2}	(exact)
Permittivity of free space	$\varepsilon_0 = (c^2\mu_0)^{-1}$	$8.854187\ldots \times 10^{-12}$ C^2·J^{-1}· m$^{-'}$	(exact)
Gravitational constant	G	6.673×10^{-11} m^3·kg^{-1}·s^{-2}	1500
Elementary charge (charge on proton or electron)	e	$1.602\,176\,462 \times 10^{-19}$C	0.039
Planck constant	h	$6.626\,068\,76 \times 10^{-34}$ J· s	0.078
	$\hbar = h/2\pi$	$1.054\,571\,596 \times 10^{-34}$ J· s	0.078
Avogadro constant	N_A	$6.022\,141\,99 \times 10^{23}$ mol^{-1}	0.079
Faraday constant	$F = N_A e$	$9.648\,534\,15 \times 10^4$ C· mol^{-1}	0.04
Electron mass	m_e	$9.109\,381\,88 \times 10^{-31}$ kg	0.079
Bohr radius	$a_0 = 4\pi\varepsilon_0\hbar^2/m_e e^2$	$5.291\,772\,083 \times 10^{-11}$ m	0.0037
Atomic mass unit (amu)	$m_u = m(^{12}\text{C})/12$	$1.660\,538\,73 \times 10^{-27}$ kg	0.079
Proton mass	m_p	$1.672\,621\,58 \times 10^{-27}$ kg	0.079
Neutron mass	m_n	$1.674\,927\,16 \times 10^{-27}$ kg	0.079
Deuteron mass	m_d	$3.343\,583\,09 \times 10^{-27}$ kg	0.079
Electron magnetic moment	μ_e	$-9.28\,476\,362 \times 10^{-24}$ J·T^{-1}	0.04
Proton magnetic moment	$\mu_p = \gamma\hbar/2$	$1.410\,606\,633 \times 10^{-26}$ J· T^{-1}	0.041
Proton gyromagnetic ratio	γ_p	$2.675\,222\,12 \times 10^8$rad·s^{-1}·T^{-1}	0.041
Molar gas constant	R	$8.314\,472$ J· mol^{-1}· K^{-1}	1.7
Boltzmann constant	$k_B = R/N_A$	$1.380\,650\,3 \times 10^{-23}$ J·K^{-1}	1.7
Molar volume,			
$T_0 = 273.15$ K, $p_0 = 10^5$ Pa	$V_m RT_0/p_0$	$22.710\,981 \times 10^{-3}$ m^3· mol^{-1}	1.7
$T_0 = 273.15$ K, $p_0 = 1$ atm		$22.413\,996 \times 10^{-3}$ m^3· mol^{-1}	1.7
(1 atm = 101 325 Pa)			
Stefan-Boltzmann constant	$\sigma = (\pi^2/60)k_B^4/\hbar^3c^2$	$5.670\,400 \times 10^{-8}$ W· m^{-2}· K^{-4}	7

Common Conversions:

Energy:	$1 \text{ kJ} \cdot \text{mol}^{-1} = 1.660540 \times 10^{-21} \text{ J}$
1 electron volt (eV):	$1.602177 \times 10^{-19} \text{ J} = 96.4853 \text{ kJ} \cdot \text{mol}^{-1}$
1 calorie (cal):	4.184 J
Length:	$1 \text{ Angstrom (Å)} = 10^{-10} \text{ m} = 100 \text{ pm}$
Mass:	$1 \text{ atomic mass unit (amu)} = 1.6605402 \times 10^{-27} \text{ kg}$
Pressure:	$1 \text{ atmosphere (atm)} = 101.325 \text{ Pa}$
1 bar:	$100{,}000 \text{ Pa}$
1 torr:	$1/760 \text{ atm}; 133.32 \text{ Pa}$
Temperature:	degrees Celsius (°C) = K -273.15
Volume:	$1 \text{ liter (L)} = .001 \text{ m}^3$

TABLE A.1 ▶ Masses of Common Isotopes (in amu):

^{1}H	1.0078250	^{2}H (also called D)	2.0141018
^{12}C	12 (exactly)	^{13}C	13.0033548
^{14}N	14.0030740	^{15}N	15.00010897
^{16}O	15.9949146		
^{19}F	18.9984032		
^{23}Na	22.989768		
^{35}Cl	34.9688527	^{37}Cl	36.9659026
^{39}K	38.963707		
^{79}Br	78.918336	^{81}Br	80.916289
^{127}I	126.904473		

Integral Formulas: Indefinite and Definite

TABLE B.1 ► Indefinite Integrals[*]

$\int x^n \, dx = x^{n+1}/(n+1), \, n \neq -1$	B-1
$\int \frac{dx}{x} = \log x$	B-2
$\int e^{ax} \, dx = e^{ax}/a$	B-3
$\int \log x \, dx = x \log x - x$	B-4
$\int (\sin ax) \, dx = -\frac{1}{a} \cos ax$	B-5
$\int (\cos ax) \, dx = \frac{1}{a} \sin ax$	B-6
$\int (\tan ax) \, dx = -\frac{1}{a} \log \cos ax$	B-7
$\int (\sin^2 ax) \, dx = -\frac{1}{2a} \cos ax \sin ax + \frac{1}{2}x = \frac{1}{2}x - \frac{1}{4a} \sin 2ax$	B-8
$\int (\cos^2 ax) \, dx = \frac{1}{2a} \sin ax \cos ax + \frac{1}{2}x = \frac{1}{2}x + \frac{1}{4a} \sin 2ax$	B-9
$\int (\sin mx)(\sin nx) \, dx = \dfrac{\sin(m-n)x}{2(m-n)} - \dfrac{\sin(m+n)x}{2(m+n)}, \, (m^2 \neq n^2)$	B-10
$\int (\cos mx)(\cos nx) \, dx = \dfrac{\sin(m-n)x}{2(m-n)} + \dfrac{\sin(m+n)x}{2(m+n)}, \, (m^2 \neq n^2)$	B-11
$\int (\sin mx)(\cos nx) \, dx = -\dfrac{\cos(m-n)x}{2(m-n)} - \dfrac{\cos(m+n)x}{2(m+n)}, \, (m^2 \neq n^2)$	B-12

[*]For each equation, an arbitrary constant C may be added to the result.

Indefinite Integrals (cont.)

$$\int (\sin^2 ax)(\cos^2 ax)\,dx = -\frac{1}{32a}\sin 4ax + \frac{x}{8} \qquad \text{B-13}$$

$$\int (\sin ax)(\cos^m ax)\,dx = -\frac{\cos^{m+1} ax}{(m+1)a} \qquad \text{B-14}$$

$$\int (\sin^m ax)(\cos ax)\,dx = \frac{\sin^{m+1} ax}{(m+1)a} \qquad \text{B-15}$$

$$\int (\log x)\,dx = x\log x - x \qquad \text{B-16}$$

$$\int xe^{ax}\,dx = \frac{e^{ax}}{a^2}(ax-1) \qquad \text{B-17}$$

TABLE B.2 ► Useful Definite Integrals: Gaussian and Related Integrals

In the expressions below, a, L, and σ are real constants.

$$\int_{x=-\infty}^{x=\infty} e^{-x^2/2\sigma^2}\,dx = \sigma\sqrt{2\pi} \qquad \text{B-18}$$

$$\int_{x=-\infty}^{x=\infty} |x|\,e^{-x^2/2\sigma^2}\,dx = 2\sigma^2 \qquad \text{B-19}$$

$$\int_{x=-\infty}^{x=\infty} x^2 e^{-x^2/2\sigma^2}\,dx = \sigma^3\sqrt{2\pi} \qquad \text{B-20}$$

$$\int_0^\infty x^{2n} e^{-ax^2}\,dx = \frac{1\cdot 3\cdot 5\cdots(2n-1)}{2^{n+1}a^n}\left(\frac{\pi}{a}\right)^{1/2} \quad (n\text{ positive integer}) \qquad \text{B-21}$$

$$\int_0^\infty x^{2n+1} e^{-ax^2}\,dx = \frac{n!}{2a^{n+1}} \quad (a>0) \qquad \text{B-22}$$

TABLE B.3 ▶ **The Area under a Gaussian Curve between Different Limits**

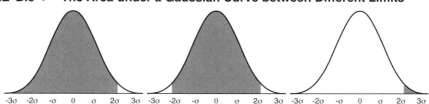

z	less than $+z\sigma$	between $\pm z\sigma$	greater than $+z\sigma$
0	0.500	0.000	0.500
0.5	0.695	0.383	0.305
1.0	0.841	0.682	0.159
1.282	0.900	0.800	0.100
1.5	0.9332	0.866	0.0668
1.645	0.950	0.900	0.050
1.960	0.975	0.950	0.025
2.0	0.9772	0.954	0.0228
2.326	0.990	0.980	0.010
2.5	0.9938	0.988	0.00621
3.0	0.9986	0.997	0.00135
3.5	0.9998	0.999	0.000233
4.0	0.9999609	0.9999	0.0000391
5.0	$1-2.87 \times 10^{-7}$	$1-5.74 \times 10^{-7}$	2.87×10^{-7}
10.0	$1-7.62 \times 10^{-24}$	$1-1.52 \times 10^{-23}$	7.62×10^{-24}

When $z \gg 1$, the following approximate formula is useful:

$$\frac{1}{\sigma\sqrt{2\pi}} \int_{z\sigma}^{\infty} e^{-x^2/2\sigma^2}\, dx \approx \frac{e^{-(z^2/2)}}{z\sqrt{2\pi}} \quad (z \gg 1)$$

Additional Readings

It would be possible to compile a list of additional readings for the range of subjects in this text which would be as long as the text itself. Instead, I choose to err on the side of brevity. The books listed below provide general information on most of these subjects. Links to other, more specialized texts are included on the Web page.

The original source of most of the quotations in the text (if it is not listed with the quote) can be found in Alan McCay, *A Dictionary of Scientific Quotations* (Institute of Physics Publishing, Bristol, 1992).

1. *Handbook of Chemistry and Physics* (D. R. Lide, editor; CRC Press, Boca Raton, FL; published biannually). This book is probably the reference work which is most universally owned by physicists and chemists. Most of the information never goes out of date, and it is often possible to purchase a previous edition at a large discount.

2. I. S. Gradshetyn and I. M. Ryzhik, *Tables of Integrals, Series and Products* (Academic Press, New York, 1980) is one of many reference tabulations of integrals and derivatives.

3. *A Physicist's Desk Reference* (H. L. Anderson, editor; American Institute of Physics, New York, 1989) collects useful formulas, constants, and facts from all branches of physics, from the undergraduate to the graduate level. It is also far more compact (and less expensive) than reference [1] above.

4. J. C. Polkinghorne, *The Quantum World* (Princeton University Press, Princeton, NJ, 1989) has a wonderful treatment of the philosophical consequences of quantum mechanics.

5. Thomas S. Kuhn, *Black-Body Theory and the Quantum Discontinuity* (Oxford University Press, New York, 1978) provides an overview, with references to the original works, of the beginnings of the quantum theory.

6. Gordon Barrow, *The Structure of Molecules* (W. A. Benjamin, New York, 1963), despite its age, is still an excellent introductory treatment of molecular spectroscopy at this level.

7. A good starting point for understanding laser design and chemical applications is D. L. Andrews, *Lasers in Chemistry* (Springer-Verlag, Berlin, 1990).

8. Roland Omns, *Quantum Philosophy: Understanding and Interpreting Contemporary Science* (Princeton University Press, 1999) presents an excellent treatment of the philosophical consequences of quantum mechanics.

This book is written at a far lower level than any physical chemistry text, but most of those books also cover all of the material presented here. In addition, an excursion through the catalog of a good college or university chemistry library is recommended.

Answers

Chapter 1

1-1. The volume is 22.414L

1-3. First find the number of grams of silicon in the unit cell, by multiplying the density of silicon by the volume of the unit cell. Then use the atomic weight of silicon to determine how many moles of silicon are in the unit cell. This is eight atoms.

1-5. 1.47×10^{-4} moles per liter

1-7. 1.05×10^{-4} moles per liter

1-9. 1.39×10^{-8} moles per liter. The assumption is that iodide from the dissolved lead iodide does not affect the concentration of iodide in solutions—an excellent approximation in this case.

1-11. $r = 1, \theta = 0, \phi = \pi/2$

1-13. a) plane;   c) plane;   e) cone

1-15. a) $70.5°$

1-17. domain $[-1, 1]$, range $[-\pi/2, \pi/2]$

1-19. $\log 50 = \log(100/2) = \log 100 - \log 2 = 2 - .301 = 1.699$

Chapter 2

2-1. $dy/dx = -2 - 2x$

2-3. a) $df(x)/dx = 2 \sin x \cos x$
b) $df(x)/dx = 1/x$

2-5. $\ln(1 + x) \approx x$

2-7. $\ln(1 + x) = x - x^2/2 + x^3/3 + \ldots - x^{2n}/(2n) + x^{(2n+1)}/(2n + 1) + \ldots$

2-9. $\int\limits_{x=0}^{x=\pi/2} \sin x \, dx = (-\cos x)|_{x=0}^{x=\pi/2} = (0) - (-1) = 1$

$\int\limits_{x=0}^{x=1} e^{2x} \, dx = (e^{2x}/2)|_{x=0}^{x=1} = e^2/2 - 1/2$

2-11. $\int\limits_{0}^{\infty} e^{-ax^2} \, dx = \left(\dfrac{\pi}{4a}\right)^{1/2}$

2-13. $\dfrac{d[C_4H_6](t)}{dt} = -k\{[C_4H_6](t)\}$; half-life $= \dfrac{1}{k[C_4H_6](t = 0)}$

Chapter 3

3-1. The gravitational force between a proton and an electron is about 4×10^{-40} of the Coulombic force.

3-3. Escape velocity is 11.179 km·s^{-1}

3-5. $x(t) = L\cos(\omega t)$; $v(t) = -\omega L \sin(\omega t)$; $\omega = \sqrt{k/m}$
$K = mv^2/2 = m\omega^2 L^2 \sin^2(\omega t)/2 = kL^2 \sin^2(\omega t)/2$ (since $\omega^2 = k/m$)
$U = kx^2/2 = kL^2 \cos^2(\omega t)/2$
$K + U = kL^2/2$

3-7. This spacing gives $d = 8.33 \times 10^{-7}$ m. For $\lambda = 4.88 \times 10^{-7}$ m and $N = 1$, the diffraction equation gives $\theta = 0.6259$ radians. After one meter this beam is deflected by 1 m·$(\tan(.6259)) = 0.723$ m. For $\lambda = 5.14 \times 10^{-7}$ m, the diffraction equation gives $\theta = 0.6650$ radians (a larger angle at longer wavelength), and after one meter the beam is deflected by 0.784 m. So the beams are separated by 61 mm.

3-9. Near-grazing incidence gives much higher resolution. A common lecture demonstration is to use the lines on a ruler at near-grazing incidence to measure the wavelength of a helium-neon laser (0.633 microns).

3-11. The pressure is 133 kPa. The pressure exerted by a 760 mm column will work out to be exactly one atmosphere.

3-13. The vibrational frequencies of H_2, HD, and D_2 are 132, 114, and 93 THz respectively.

3-15. The reaction of H_2 and Cl_2 is highly exothermic, releasing 185 kJ per mole.

Chapter 4

4-1. The probability of getting 100 heads is 2^{-100}
The probability of getting 99 heads and one tail is $100 \cdot 2^{-100}$
The probability of getting 98 heads and two tails is $(100 \cdot 99/2) \cdot 2^{-100}$

The probability of getting 97 heads and three tails is $(100 \cdot 99 \cdot 98/6) \cdot 2^{-100}$
The probability of getting 96 heads and four tails is $(100 \cdot 99 \cdot 98 \cdot 97/24) \cdot 2^{-100}$
The probability of getting 95 heads and five tails is $(100 \cdot 99 \cdot 98 \cdot 97 \cdot 96/120) \cdot 2^{-100}$
The sum of all of these numbers is 6.26×10^{-23}.

4-3. Either of these is the probability of $|M| \geq 4\sigma$, which is 3.91×10^{-5}.

4-5. c) Using the formulas presented in the problem, you should calculate a mean of 100.4283, a variance of 0.93105, and 95% confidence limits of 0.97724. So you would report the mean as 100.4 ± 1.0 mM, and you cannot say with 95% confidence that the average is above 100 mM.

4-7. From the Boltzmann distribution, the ratio of pressures (assuming constant temperature) should be $e^{-mgh/k_B T}$. Since $mg/k_B T = 1.26 \times 10^{-4}$ m^{-1} (see text), at a height of 1500 m, $mgh/k_B T = .189$, and the pressure is predicted to be about 83% of the pressure at sea level.

4-9. If we assume that boys and girls have equal birth probabilities, for two children there are four equally likely outcomes: boy-boy, boy-girl, girl-boy, and girl-girl. Three of these fit Mary's description (at least one boy) so the chance that she has two boys is 1/3. Two of them fit Jane's description (the first child is a boy) so the chance that she has two boys is 1/2.

4-11. $K = 4$

4-13. If the errors in A and B are random and uncorrelated with one another, sometimes A will be larger than its true value and B will be smaller, or A will be smaller than its true value and B will be larger. In either case the product AB is then accidentally closer to the true value than one might expect. In the product A·A, this accidental cancellation of errors does not happen.

Chapter 5

5-1. These numbers give M_2O_3, with atomic weight 56 g· mol^{-1}.

5-3. Neither the warm engine nor the surrounding grass gives off much visible light by blackbody radiation, but both radiate in the infrared. Assume the grass has a temperature of 290K and the engine has a temperature of 320K. For perfect blackbodies, the warm engine would radiate $(320/290)^4 = 1.48$ times as much energy.

5-5. $c_v = \dfrac{d\langle E \rangle}{dt} = k_B \dfrac{e^{h\nu/k_B T}(h\nu/k_B T)^2}{\left(e^{h\nu/k_B T} - 1\right)^2}$, so c_v approaches k_B at very high temperatures, and is very small at low temperatures.

5-7. $\lambda = 650$ nm implies $E = hc/\lambda = 3.05 \times 10^{-19}$ Joules per photon, or 184 kJ· mol^{-1}. At 5 mW average power (.005 J/s) it would take about 10, 200 hours to produce one einstein.

5-9. The relation $\Delta E \Delta t \geq h/4$ gives an uncertainty in the energy, and Einstein's relation $E = mc^2$ converts this into a mass. Substituting in $\Delta t = 12$ min (720 s)
gives $\Delta m \geq 10^{-54}$ kg—not a serious limitation.

5-11. a) $K = 2.18 \times 10^{-18}$ J.
b) The momentum vector of length $(2mK)^{1/2}$ is random in direction, so $\Delta p \approx$
$(2mK)^{1/2} = 1.99 \times 10^{-24}$ kg·m·s^{-1}.
c) Plugging this into the Uncertainty Principle relationship gives $\Delta x \geq 8.3 \times 10^{-11}$
m, which is greater than a_0 itself. This is only a crude calculation (replacing a
three-dimensional distribution with a one-dimensional uncertainty) but the result
is essentially correct—the electron must be delocalized over a wide region of space
(as we will see in Chapter 6).

5-13. Note from Equation 5.48 that the spin up state has the lower energy. The ratio of
populations between the two states is

$$\frac{N_\alpha}{N_\beta} = e^{h \cdot (426 \text{ MHz})/k_B T} = e^{6.8 \times 10^{-5}} = 1.000068$$

so the fraction of population in the higher state (β) is

$$\frac{N_\beta}{N_\alpha + N_\beta} = \frac{1}{N_\alpha/N_\beta + 1} = \frac{1}{1.000068 + 1} = .499983$$

The populations of the two states are very nearly equal.

5-15. $E_n = \dfrac{n^2 h^2}{8mL^2}$

Chapter 6

6-1. a) $\left| e^{i\theta} \right| = |\cos\theta + i\sin\theta| = \sqrt{\cos^2\theta + \sin^2\theta} = 1$

6-3. Since $\Psi(t) = e^{-iEt/\hbar}\Psi(0)$, we can take the complex conjugate to show $\Psi^*(t) =$
$e^{+iEt/\hbar}\Psi^*(0)$. The probability distribution at any time t is given by $P(t) =$
$\Psi(t)\Psi^*(t) = e^{-iEt/\hbar}e^{+iEt/\hbar}\Psi(0)\Psi^*(0) = \Psi(0)\Psi^*(0) = P(0)$, so the probability distribution is independent of time.

6-5. Equation [B-8] in Appendix B, $\int (\sin^2 ax)\,dx = \dfrac{1}{2}x - \dfrac{1}{4a}\sin 2ax$, can be used
with $a = n\pi/L$. This integral is made even simpler by realizing that $\sin(2ax) =$
$\sin(2\pi x/L)$ vanishes at the upper and lower limits of the integral, so in fact the
integral from $x = 0$ to $x = L$ is equal to $L/2$. The normalization constant must
then be $(2/L)^{1/2}$.

6-7. Your graph should give a wavefunction localized in the right side of the box—the
mirror image about $x = L/2$ of Figure 6.3.

6-9. Your graph should be symmetric about $x = L/2$, so $\langle x \rangle = L/2$, and $\langle p \rangle = 0$
because the wavefunction is real.

6-11. b) $E = \hbar\omega_0/2$

6-13. As with many of the paradoxical results of quantum mechanics, the Uncertainty Principle comes to the rescue. You cannot localize the position of the electron inside the nucleus (very small Δx) without creating a huge uncertainty in the momentum, and thus losing any knowledge of the orbital you are in.

Chapter 7

7-1. The momentum and kinetic energy conservation equations are:

$$6.6 \times 10^{-24} \text{ kg} \cdot \text{m} \cdot \text{s}^{-1} = (6.6 \times 10^{-27} \text{ kg})v_{\text{He,final}} + (1 \text{ kg})v_{\text{wall,final}}$$
$$3.3 \times 10^{-21} \text{ J} = (6.6 \times 10^{-27} \text{ kg})(v_{\text{He,final}})^2/2 + (1 \text{ kg})(v_{\text{wall,final}})^2/2$$

Rearrange the first equation to give:

$$v_{\text{He,final}} = 1000 \text{ m} \cdot \text{s}^{-1} - (1.52 \times 10^{26})v_{\text{wall,final}}$$

and substitute into the second equation to give a quadratic equation:

$$3.3 \times 10^{-21} \text{ J} = (6.6 \times 10^{-27} \text{ kg})(1000 \text{ m} \cdot \text{s}^{-1} - (1.52 \times 10^{26})v_{\text{wall,final}})^2/2$$
$$+(1 \text{ kg})(v_{\text{wall,final}})^2/2$$

or

$$(7.6 \times 10^{25} \text{ kg})(v_{\text{wall,final}})^2 - (1000 \text{ kg} \cdot \text{m} \cdot \text{s}^{-1})v_{\text{wall,final}} = 0$$

This has two solutions:

$$v_{\text{wall,final}} = 0, v_{\text{He,final}} = 1000 \text{ m} \cdot \text{s}^{-1} \text{ (the initial condition)}$$
$$v_{\text{wall,final}} = 1.31 \times 10^{-23} \text{ m} \cdot \text{s}^{-1}, v_{\text{He,final}} = -1000 \text{ m} \cdot \text{s}^{-1} \text{ (the final condition)}$$

7-3. (a) The mean free path stays constant. The distance in any direction to the nearest obstacle is independent of temperature.
(b) The mean time between collisions decreases by a factor of $\sqrt{2}$, because the speed goes up by that factor and the mean free path is constant.
(c) The diffusion constant increases by $\sqrt{2}$, because the collision frequency (the inverse of the mean time between collisions) increases by $\sqrt{2}$. Note that this is not what you would predict by just doubling the temperature in Equation 7.41— the assumption here was that the temperature was doubled while keeping N/V constant, so the pressure doubles as well. If you double both the temperature and the pressure in Equation 7.41 you get the same answer.

7-5. The integral for evaluating $\langle s^2 \rangle$ has the form $\int_0^\infty x^4 e^{-ax^2} dx \ (= 3\sqrt{\pi}/8a^{5/2})$; the integral for evaluating $\langle s \rangle$ has the form $\int_0^\infty x^3 e^{-ax^2} dx \ (= (2a^2)^{-1})$. Both integrals can be found in the more general form in Appendix B.

7-7. $PV/nRT = 1 + B(t)(n/V) + \ldots$; the problem specifies that at $T = 273K$, $P = 1$ atm and $n = 1$ mole, $PV/nRT = (22.260/22.41410) = .9931$ instead of 1. So

$$B(T) * (1/22.260) = .9931 - 1 = .0069$$
$$B(T) = -0.153 \text{ liters per mole}$$

Chapter 8

8-1. 128 pm (see Table 3.2)

8-3. $\nu = 6.84$ GHz

8-5. $B = 2.01$ cm^{-1}; $\omega_e = 2359.61$ cm^{-1}. Note that neither of these values can be obtained from infrared or microwave spectra, because the molecule has no dipole moment—infrared or microwave radiation will not induce transitions between the different vibrational and rotational levels. They are obtained from electronic spectra.

8-7. Tetracene looks orange-red. It gives off blue-green fluorescence.

8-9. For the particle in a box we have

$$E = \frac{(n_x^2 + n_y^2 + n_z^2)/h^2}{8mV^{2/3}}; \quad P = -\frac{dE}{dV} = \frac{+2}{3}\frac{(n_x^2 + n_y^2 + n_z^2)h^2}{8mV^{5/3}} = \frac{2E}{3V}$$

8-11. The lines are at multiples of 13.102 cm^{-1}.

8-13. The vibrational frequencies are lower, so it requires still higher values of Δv to get absorption in the visible. These transitions are still weaker, so the prediction is that while red light is still absorbed more than blue (lower Δv to get to red), all transitions are weaker and the pool would be more nearly colorless.

Index